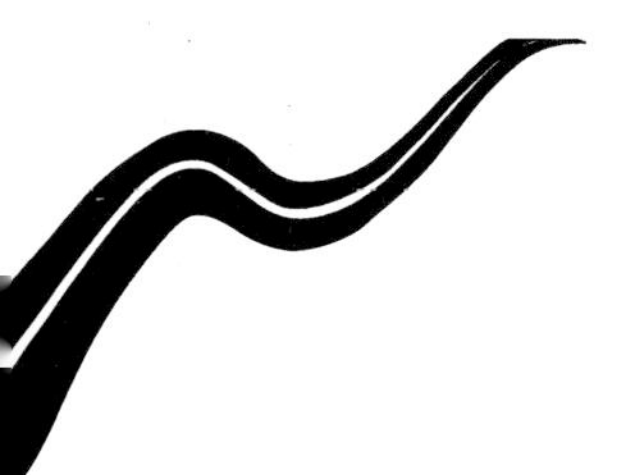

Supplement to the
AASHTO Guide For

Design of Pavement Structures

Part II, - Rigid Pavement Design & Rigid Pavement Joint Design

Published by the
American Association of State Highway
and Transportation Officials

ISBN 1-56051-078-1

AMERICAN ASSOCIATION OF STATE HIGHWAY AND TRANSPORTATION OFFICIALS

EXECUTIVE COMMITTEE
1996-1997

VOTING MEMBERS

Officers:

PRESIDENT: Darrel Rensink, Iowa

VICE PRESIDENT: David Winstead, Maryland

SECRETARY-TREASURER: Clyde E. Pyers, Maryland

Regional Representatives:

REGION I: Carlos I. Pesquera, Puerto Rico

REGION II: Robert L. Robinson, Mississippi

REGION III: Robert A. Welke, Michigan

REGION IV: Marshall W. Moore, North Dakota

NON-VOTING MEMBERS

Immediate Past President: Wm. G. Burnett, P.E., Texas

Executive Director: Francis B. Francois, Washington, D. C.

**AMERICAN ASSOCIATION OF STATE HIGHWAY
AND TRANSPORTATION OFFICIALS**

JOINT TASK FORCE ON PAVEMENTS

Officers:

> CHAIRMAN: Gary Carver, Wyoming
> VICE CHAIRMAN: Robert L. Walters, Arkansas
> SECRETARY: Paul Teng, FHWA

Regional Representatives:

Region 1

Connecticut	Charles E. Dougan
New York	Wes Yang
Pennsylvania	Danny Dawood
Port Authority of NY and NJ	Harry Schmerl
FHWA	Paul Yeng

Region 2

Arkansas	Robert L. Walters
Kentucky	Gary Sharpe
Louisiana	J.B. Esnard, Jr.
North Carolina	Thomas M. Hearne, Jr.

Region 3

Missouri	Jay F. Bledsoe
Ohio	Aric Morse
Wisconsin	Stephen F. Shober

Region 4

Arizona	George Way
California	Kevin Herritt
Colorado	Steve Horton
Texas	Ken Fultz
Washington	Linda Pierce
Wyoming	F.M. (Rick) Harvey

Representing

Transportation Research Board	Amir N. Hanna
Standing Committee on Planning	Fred Van Kirk
Subcommittees on Construction and Maintenance	Dean M. Testa
Subcommittee on Materials	Larry Engbrecht
Standing Committee on Aviation	Craig Smith

Preface

The Joint Task Force on Pavements—composed of members from the Highway Subcommittee on Design, one member each from the Highway Subcommittees on Materials, Construction, and Maintenance, and one from the Standing Committee on Planning—has developed this Supplement to the *AASHTO Guide for the Design of Pavement Structures*. This Supplement includes alternative design procedures that can be used in place of or in conjunction with Part II, Section 3.2 "Rigid Pavement Design" and Section 3.3 "Rigid Pavement Joint Design."

The development of these alternative design procedures began with AASHTO's NCHRP Project 1-30. This project was to utilize new data available as a result of the LTPP database to incorporate new data on loss-of-support for use in the design of rigid pavements and overlays and to improve the selection of k values. The document was then validated and revised under an FHWA contract and then finally developed into design procedures for use by AASHTO. These alternative design procedures and a "Rigid Pavement Design Example" are included herein for those who would like to make use of the LTPP data in rigid pavement design.

This supplement has been ballotted and approved for publication by AASHTO's Standing Committee on Highways.

SUPPLEMENTAL VERSION OF THE AASHTO GUIDE, PART II, SECTION 3.2 RIGID PAVEMENT DESIGN FOR DESIGN OF PAVEMENT STRUCTURES AND SECTION 3.3 RIGID PAVEMENT JOINT DESIGN

This supplement has been prepared as an alternate method for rigid pavement design, in the form of an addendum to the current AASHTO Guide. It contains the recommendations from NCHRP 1-30, modified based on the results of the verification study conducted using the LTPP database.

3.2 RIGID PAVEMENT DESIGN

This section describes the design for Portland cement concrete pavements, including jointed plain (JPCP), jointed reinforced (JRCP), and continuously reinforced (CRCP). As in the design for flexible pavements, it is assumed that these pavements will carry traffic levels in excess of 70,000 18-kip [80-kN] (rigid pavement) ESALs over the performance period. Examples of use of this rigid pavement design procedure are presented at the end of this supplement.

Design of Different Types of Concrete Pavement. The JPCP design concept is to provide a sufficient slab thickness and joint spacing to minimize the development of transverse cracking. The JRCP and CRCP design concepts provide sufficient slab thickness and reinforcement to hold very tight the transverse cracks that form so that aggregate interlock will be maintained. The thickness of the design model upon which this guide is based was developed and validated specifically for JPCP, for which joint spacing is one of the important required design inputs affecting thermal curling stresses and, thus, transverse cracking. A proper selection of slab thickness and joint spacing is required to control the development of transverse cracking for a given climate, base, and subgrade. JRCP has much longer joint spacing and CRCP has no joints, and the transverse cracks that eventually form in these types of pavements must be held tight by sufficient steel reinforcement.

The use of this design method to determine an appropriate slab thickness for JRCP or CRCP requires the selection of an input "hypothetical" joint spacing. Research using the LTPP database has shown that the following input values of joint spacing will result in reasonable design thicknesses using this design method.

JPCP: Actual joint spacing, ft.

JRCP: Actual joint spacing if less than 30 ft [9 m], or 30 ft maximum (use this value only to obtain slab design thickness).

CRCP: 15 ft [4.6 m] (use this hypothetical value only to obtain slab design thickness).

1

Load Transfer at Joints. The AASHTO design procedure is based on the AASHO Road Test pavement performance algorithm that was extended to include additional design features. Inherent in the use of the AASHTO procedure is the use of dowels at transverse joints. Joint faulting was not a distress manifestation at the Road Test due to the adequacy of the dowel design. A faulting design check is provided for doweled joints to ensure that the dowels are sized properly. If a significant faulting problem is expected, an increase in dowel diameter or other design change may be warranted. The non-doweled faulting check was developed using more recent measurements of field data.

If the designer wishes to consider undoweled joints, a design check for faulting is provided. If the faulting check indicates inadequate load transfer, design modifications such as the use of dowels or changes in base type, drainage, and joint spacing may be made.

In addition, if the designer wishes to consider undoweled joints, a design check is also made for critical stresses due to axle loads applied near the transverse joint, along with a negative thermal gradient, creating a corner loading situation that would lead to premature cracking. If this check shows a potential problem, design modifications such as the use of dowels, increased slab thickness, or changes in base type may be made.

3.2.1 Develop Effective Modulus of Subgrade Reaction (k-Value)

The modulus of subgrade reaction (k-value) is defined as that measured or estimated on top of the finished roadbed soil or embankment upon which the base course and/or concrete slab will eventually be constructed. The k-value represents the subgrade (and embankment, if present); it does not represent the base course. The base course is considered a structural layer of the pavement along with the concrete slab, and thus its thickness and modulus are important design inputs in determining the required slab thickness in Section 3.2.2.

The k-Value input defined. The elastic k-value on top of the subgrade or embankment is the required design input. The gross k-value incorporated in previous versions of the AASHTO Guide represents not only elastic deformation of the subgrade under a loading plate, but also substantial permanent deformation. Only the elastic component of this deformation is considered representative of the response of the subgrade to traffic loads on the pavement. The elastic k-value test was the main subgrade test conducted extensively at the AASHO Road Test. When the elastic k-value was used in structural analysis of the AASHO Road Test pavements, it was found that slab stresses computed with a three-dimensional finite element model were approximately equal to those measured in the field under full-scale truck axle loadings at creep speed, providing further justification for use of the elastic k-value in the design.

Steps in determining design k-value. The k-value input required for this design method is determined by the following steps, which are described in this section:

1. Select a subgrade k-value for each season, using any of the three following methods:
    (a) Correlations with soil type and other soil properties or tests.
    (b) Deflection testing and backcalculation (most highly recommended).
    (c) Plate bearing tests.

2. Determine a seasonally adjusted effective k-value.
3. Adjust the seasonal effective k-value for effects of a shallow rigid layer, if present, and/or an embankment above the natural subgrade.

Note that the AASHTO design methodology requires the mean k-value, not the lowest value measured or some other conservative value. Note also that no additional adjustment to the k-value is applied for loss of support. Substantial loss of support existed for many sections at the AASHO Road Test, which led to increased slab cracking and loss of serviceability. Therefore, the performance data, upon which the AASHO Road Test performance model is based, already reflect the effect of considerable loss of support.

Step 1. Select a Subgrade k-Value for Each Season. A season is defined as a period of time within a year, such as 3 months (i.e., spring, summer, fall, winter). The number of seasons and the length of each season by which a year is characterized depend on the climate of the pavement's location.

There are several ways to measure or estimate the subgrade elastic k-value. Procedures are provided for three methods described below—correlation methods, backcalculation methods, and plate testing methods.

Correlation Methods. Guidelines are presented for selecting an appropriate k-value based on soil classification, moisture level, density, California Bearing Ratio (CBR), or Dynamic Cone Penetrometer (DCP) data. The CBR may also be estimated from the R-value. These correlation methods are anticipated to be used routinely for design. The k-values obtained from soil type or tests correlation methods may need to be adjusted for embankment above the subgrade or a shallow rigid layer beneath the subgrade.

The k-values and correlations for cohesive soils (A-4 through A-7). The bearing capacity of cohesive soils is strongly influenced by their degree of saturation (S_r, percent), which is a function of water content (w, percent), dry density (γ, lb/ft^3), and specific gravity (G_s):

$$S_r = \frac{w}{\left(\dfrac{62.4}{\gamma}\right) - \left(\dfrac{1}{G_s}\right)} \qquad [25]$$

Recommended k-values for each fine-grained soil type as a function of degree of saturation are shown in Figure 40. Each line represents the middle of a range of reasonable values for k. For any given soil type and degree of saturation, the range of reasonable values is about $\pm$ 40 psi/in [11 kPa/mm]. A reasonable lower limit for k at 100 percent saturation is considered to be 25 psi/in [7 kPa/mm]. Thus, for example, an A-6 soil might be expected to exhibit k-values between about 180 and 260 psi/in [49 and 70 kPa/mm] at 50 percent saturation, and k-values between about 25 and 85 psi/in [7 and 23 kPa/mm] at 100 percent saturation.

Two different types of materials can be classified as A-4: predominantly silty materials (at least 75 percent passing the #200 sieve, possibly organic), and mixtures of silt, sand, and gravel (up to 64 percent retained on #200 sieve). The former may have a density between about 90 and 105 lb/ft³ [1442 and 1682 kg/m³], and a CBR between about 4 and 8. The latter may have a density between about 100 and 125 lb/ft³ [1602 and 2002 kg/m³], and a CBR between about 5 and 15. The line labeled A-4 in Figure 40 is more representative of the former group. If the material in question is A-4, but possesses the properties of the stronger subset of materials in the A-4 class, a higher k-value at any given degree of saturation (for example, along the line labeled A-7-6 in Figure 40) is appropriate.

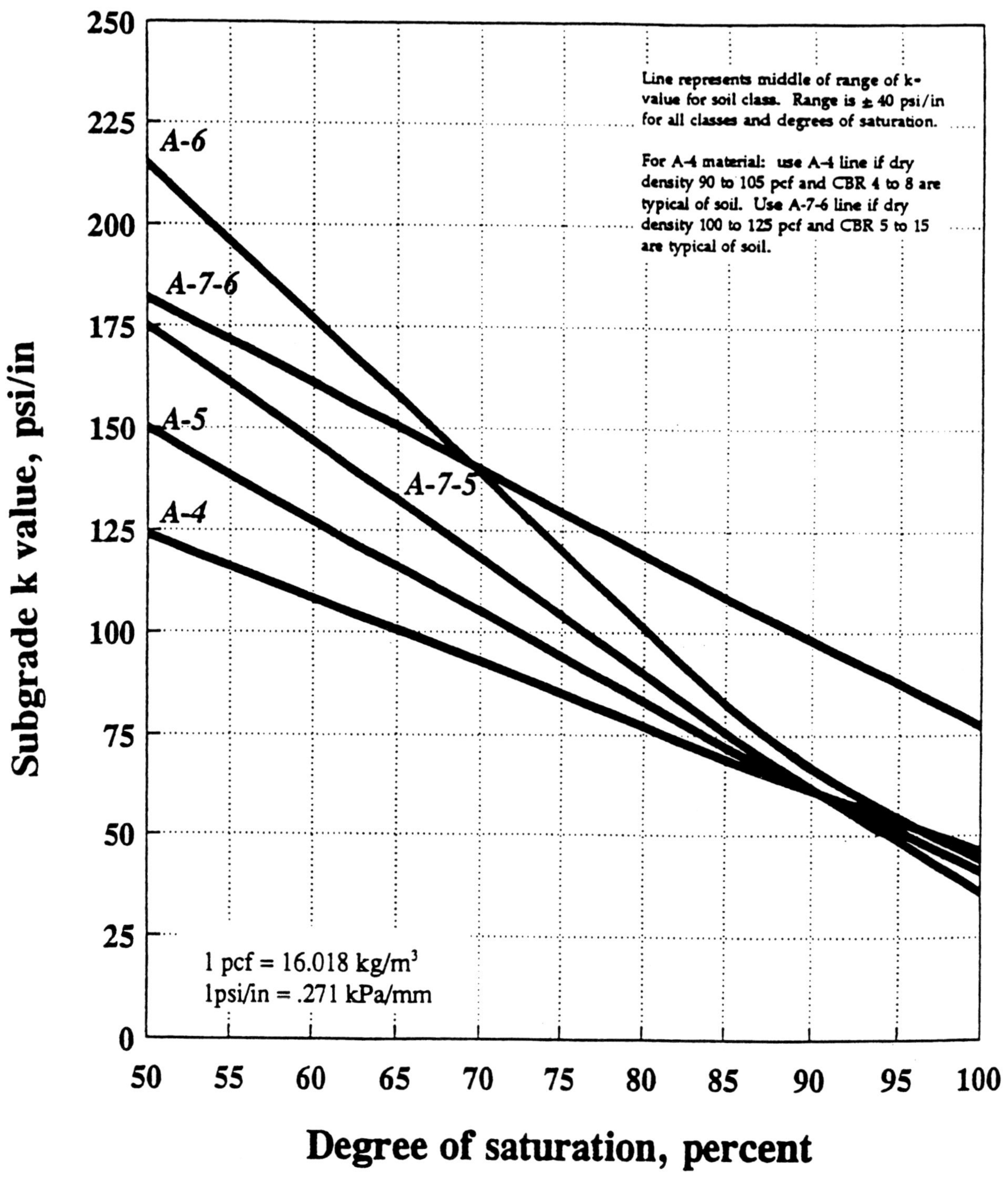

Figure 40. The k-value versus degree of saturation for cohesive soils.

Recommended k-value ranges for fine-grained soils, along with typical ranges of dry density and CBR for each soil type, are summarized in Table 11.

The k-values and correlations for cohesionless soils (A-1 and A-3). The bearing capacity of cohesionless materials is fairly insensitive to moisture variation and is predominantly a function of their void ratio and overall stress state. Recommended k-value ranges for cohesionless soils, along with typical ranges of dry density and CBR for each soil type, are summarized in Table 11.

The k-values and correlations for A-2 soils. Soils in the A-2 class are all granular materials falling between A-1 and A-3. Although it is difficult to predict the behavior of such a wide variety of materials, the available data indicate that in terms of bearing capacity, A-2 materials behave similarly to cohesionless materials of comparable density. Recommended k-value ranges for A-2 soils, along with typical ranges of dry density and CBR for each soil type, are summarized in Table 11.

Correlation of k-values to California Bearing Ratio. Figure 41 illustrates the approximate range of k-values that might be expected for a soil with a given California Bearing Ratio.

Correlation of k-values to penetration rate by Dynamic Cone Penetrometer. Figure 42 illustrates the range of k-values that might be expected for a soil with a given penetration rate (inches per blow) measured with a Dynamic Cone Penetrometer. This is a rapid hand-held testing device that can be used to quickly test dozens of locations along an alignment. The DCP can also penetrate AC surfaces and surface treatments to test the foundation below.

Assignment of k-values to seasons. Among the factors that should be considered in selecting seasonal k-values are the seasonal movement of the water table, seasonal precipitation levels, winter frost depths, number of freeze-thaw cycles, and the extent to which the subgrade will be protected from frost by embankment material. A "frozen" k may not be appropriate for winter, even in a cold climate, if the frost will not reach and remain in a substantial thickness of the subgrade throughout the winter. If it is anticipated that a substantial depth (e.g., a few feet) of the subgrade will be frozen, a k-value of 500 psi/in [135 kPa/mm] would be an appropriate "frozen" k.

The seasonal variation in degree of saturation is difficult to predict, but in locations where a water table is constantly present at a depth of less than about 10 ft [3 m], it is reasonable to expect that fine-grained subgrades will remain at least 70 to 90 percent saturated, and may be completely saturated for substantial periods in the spring. County soil reports can provide data on the position of the high-water table (i.e., the typical depth to the water table at the time of the year that it is at its highest). Unfortunately, county soil reports do not provide data on the variation in depth to the water table throughout the year.

Table 11. Recommended k-value ranges for various soil types.

AASHTO class	Description	Unified class	Dry density (lb/ft³)	CBR (percent)	k-value (psi/in)
Coarse-grained soils:					
A-1-a, well graded	gravel	GW, GP	125 - 140	60 - 80	300 - 450
A-1-a, poorly graded			120 - 130	35 - 60	300 - 400
A-1-b	coarse sand	SW	110 - 130	20 - 40	200 - 400
A-3	fine sand	SP	105 - 120	15 - 25	150 - 300
A-2 soils (granular materials with high fines):					
A-2-4, gravelly	silty gravel	GM	130 - 145	40 - 80	300 - 500
A-2-5, gravelly	silty sandy gravel				
A-2-4, sandy	silty sand	SM	120 - 135	20 - 40	300 - 400
A-2-5, sandy	silty gravelly sand				
A-2-6, gravelly	clayey gravel	GC	120 - 140	20 - 40	200 - 450
A-2-7, gravelly	clayey sandy gravel				
A-2-6, sandy	clayey sand	SC	105 - 130	10 - 20	150 - 350
A-2-7, sandy	clayey gravelly sand				
Fine-grained soils:					
A-4	silt	ML, OL	90 - 105	4 - 8	25 - 165 *
	silt/sand/ gravel mixture		100 - 125	5 - 15	40 - 220 *
A-5	poorly graded silt	MH	80 - 100	4 - 8	25 - 190 *
A-6	plastic clay	CL	100 - 125	5 - 15	25 - 255 *
A-7-5	moderately plastic elastic clay	CL, OL	90 - 125	4 - 15	25 - 215 *
A-7-6	highly plastic elastic clay	CH, OH	80 - 110	3 - 5	40 - 220 *

* k-value of fine-grained soil is highly dependent on degree of saturation. See Figure 40.

These recommended k-value ranges apply to a homogeneous soil layer at least 10 ft [3 m] thick. If an embankment layer less than 10 ft [3 m] thick exists over a softer subgrade, the k-value for the underlying soil should be estimated from this table and adjusted for the type and thickness of embankment material using Figure 43. If a layer of bedrock exists within 10 ft [3 m] of the top of the soil, the k should be adjusted using Figure 43.

1 lb/ft³ =16.018 kg/m³, 1 psi/in = 0.271 kPa/mm

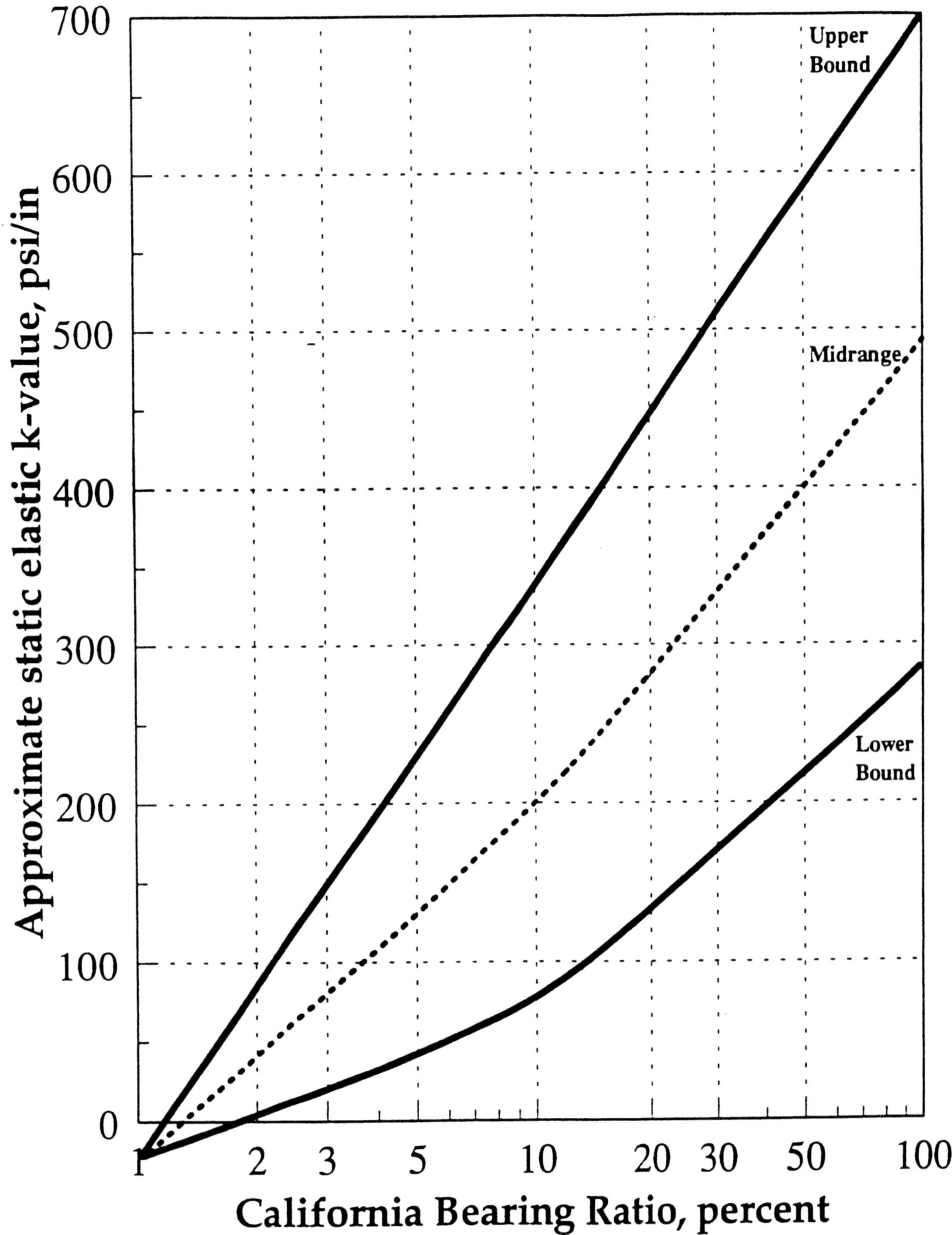

1 psi/in = 0.271 kPa/mm

Figure 41. Approximate relationship of k-value range to CBR.

7

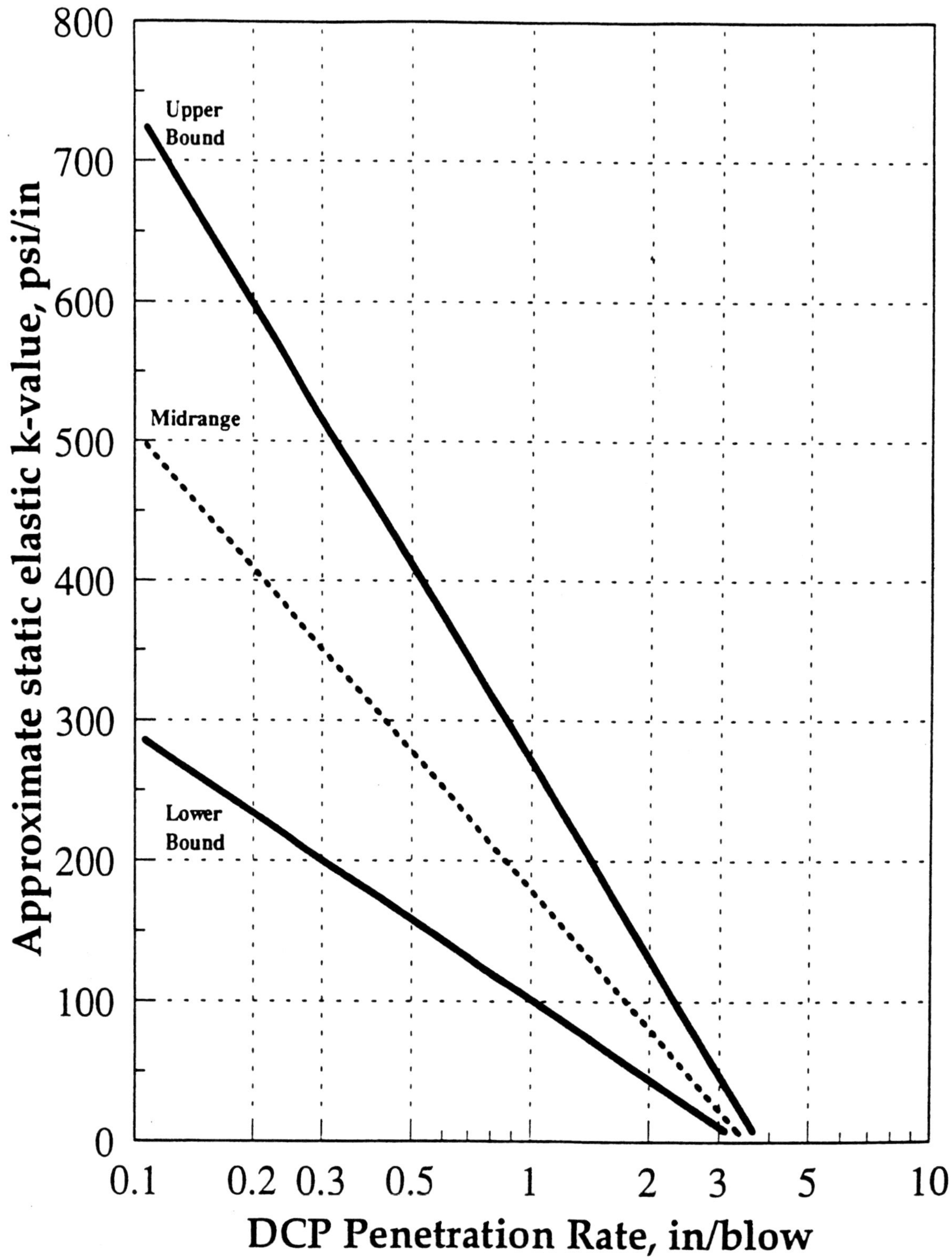

1 psi/in = 0.271 kPa/mm
1 in = 25.4 mm

Figure 42. Approximate relationship of k-value range to DCP penetration rate.

Deflection Testing and Backcalculation Methods. These methods are suitable for determining k-value for design of overlays of existing pavements, for design of a reconstructed pavement on existing alignments, or for design of similar pavements in the same general location on the same type of subgrade. An agency may also use backcalculation methods to develop correlations between nondestructive deflection testing results and subgrade types and properties. Cut and fill sections are likely to yield different k-values. No embankment or rigid layer adjustment is required for backcalculated k-values if these characteristics are similar for the pavement being tested and the pavement being designed, but backcalculated dynamic k-values do need to be reduced by a factor of approximately 2 to estimate a static elastic k-value for use in design.

An appropriate design subgrade k-value for use as an input to this design method is determined by the following steps:

1. Measure deflections on an in-service concrete or composite (AC-overlaid PCC) pavement with the same or similar subgrade as the pavement being designed.
2. Compute the appropriate AREA of each deflection basin.
3. Compute an initial estimate (assuming an infinite slab size) of the radius of relative stiffness, ℓ.
4. Compute an initial estimate (assuming an infinite slab size) of the subgrade k-value.
5. Compute adjustment factors for the maximum deflection d_0 and the initially estimated ℓ to account for the finite slab size.
6. Adjust the initially estimated k-value to account for the finite slab size.
7. Compute the mean backcalculated subgrade k-value for all of the deflection basins considered.
8. Compute the estimated mean static k-value for use in design.

These steps are described below, with the relevant equations for bare concrete and composite pavements given for each step.

Measure deflections. Measure slab deflection basins along the project at an interval sufficient to adequately assess conditions. Intervals of 100 to 1000 ft [30 to 300 m] are typical. Measure deflections with sensors located at 0, 8, 12, 18, 24, 36, and 60 in [0, 203, 305, 457, 610, 915, and 1524 mm] from the center of the load. Measure deflections in the outer wheel path. A heavy-load deflection device (e.g., Falling Weight Deflectometer) and a load magnitude of 9,000 lbf [40 kN] are recommended. ASTM D4694 and D4695 provide additional guidance on deflection testing.

Compute AREA. For a bare concrete pavement, compute the $AREA_7$ of each deflection basin using the following equation:

$$AREA_7 = 4 + 6\left(\frac{d_8}{d_0}\right) + 5\left(\frac{d_{12}}{d_0}\right) + 6\left(\frac{d_{18}}{d_0}\right) + 9\left(\frac{d_{24}}{d_0}\right) + 18\left(\frac{d_{36}}{d_0}\right) + 12\left(\frac{d_{60}}{d_0}\right) \qquad [26]$$

where d_0 = deflection in center of loading plate, inches
$\quad\ \ d_i$ = deflections at 0, 8, 12, 18, 24, 36, and 60 in [0, 203, 305, 457, 610, 915, and 1524 mm] from plate center, inches

For a composite pavement, compute the $AREA_5$ of each deflection basin using the following equation:

$$AREA_5 = 3 + 6\left(\frac{d_{18}}{d_{12}}\right) + 9\left(\frac{d_{24}}{d_{12}}\right) + 18\left(\frac{d_{36}}{d_{12}}\right) + 12\left(\frac{d_{60}}{d_{12}}\right) \qquad [27]$$

Estimate ℓ assuming an infinite slab size. The radius of relative stiffness for a bare concrete pavement (assuming an infinite slab) may be estimated using the following equation:

$$\ell_{est} = \left[\frac{\ln\left(\dfrac{60 - AREA_7}{289.708}\right)}{-0.698}\right]^{2.566} \qquad [28]$$

The radius of relative stiffness for a composite pavement (assuming an infinite slab) may be estimated using the following equation:

$$\ell_{est} = \left[\frac{\ln\left(\dfrac{48 - AREA_5}{158.40}\right)}{-0.476}\right]^{2.220} \qquad [29]$$

Estimate k assuming an infinite slab size. For a bare concrete pavement, compute an initial estimate of the k-value using the following equation:

$$k_{est} = \frac{P\, d_0^*}{d_0 \left(\ell_{est}\right)^2} \qquad [30]$$

where k = backcalculated dynamic k-value, psi/in
  P = load, lb
  d_0 = deflection measured at center of load plate, inch
  ℓ_{est} = estimated radius of relative stiffness, inches, from previous step
  d_0^* = nondimensional coefficient of deflection at center of load plate:

$$d_0^* = 0.1245\, e^{\left[-0.14707\, e^{\left(-0.07565\, \ell_{est}\right)}\right]} \qquad [31]$$

For a composite pavement, compute an initial estimate of the k-value using the following equation:

$$k_{est} = \frac{P\, d_{12}^{*}}{d_{12} \left(\ell_{est}\right)^2} \qquad [32]$$

d_{12} = deflection measured 12 in [305 mm] from center of load plate, inch
ℓ_{est} = estimated radius of relative stiffness, in, from previous step
d_{12}^{*} = nondimensional coefficient of deflection 12 in [305 mm] from center of load plate:

$$d_{12}^{*} = 0.12188\, e^{\left[-0.79432\, e^{\left(-0.07074\, \ell_{est}\right)}\right]} \qquad [33]$$

Compute adjustment factors for d_0 and ℓ for finite slab size. For both bare concrete and composite pavements, the initial estimate of ℓ is used to compute the following adjustment factors to d_0 and ℓ to account for the finite size of the slabs tested:

$$AF_{d_0} = 1 - 1.15085\, e^{\,-0.71878\left(\frac{L}{\ell_{est}}\right)^{0.80151}} \qquad [34]$$

$$AF_{\ell} = 1 - 0.89434\, e^{\,-0.61662\left(\frac{L}{\ell_{est}}\right)^{1.04831}} \qquad [35]$$

where, if the slab length is less than or equal to twice the slab width, L is the square root of the product of the slab length and width, both in inches, or if the slab length is greater than twice the width, L is the product of the square root of two and the slab length in inches:

$$if\ L_l \leq 2 * L_w, \quad L = \sqrt{L_l\, L_w}$$
$$\qquad [36]$$
$$if\ L_l > 2 * L_w, \quad L = \sqrt{2} * L_l$$

Adjust k for finite slab size. For both bare concrete and composite pavements, adjust the initially estimated k-value using the following equation:

$$k = \frac{k_{est}}{AF_{\ell}^2\ AF_{d_0}} \qquad [37]$$

Compute mean dynamic k-value. Exclude from the calculation of the mean k-value any unrealistic values (i.e., less than 50 psi/in [14 kPa/mm] or greater than 1500 psi/in [407 kPa/mm]), as well as any individual values that appear to be significantly out of line with the rest of the values.

Compute the estimated mean static k-value for design. Divide the mean dynamic k-value by two to estimate the mean static k-value for design.

A blank worksheet for computation of k from deflection data and example computations of k from deflection basins measured on two pavements, one bare concrete and the other composite, are given in Table 12.

Seasonal variation in backcalculated k-values. The design k-value determined from backcalculation as described above represents the k-value for the season in which the deflection testing was conducted. An agency may wish to conduct deflection testing on selected projects in different seasons of the year to assess the seasonal variation in backcalculated k-values for different types of subgrades.

Plate Bearing Test Methods. The subgrade or embankment k-value may be determined from either of two types of plate bearing tests: repetitive static plate loading (AASHTO T221, ASTM D1195) or nonrepetitive static plate loading (AASHTO T222, ASTM D1196). These test methods were developed for a variety of purposes, and do not provide explicit guidance on the determination of the required k-value input to the design procedure described here.

For the purpose of concrete pavement design, the recommended subgrade input parameter is the static elastic k-value. This may be determined from either a repetitive or nonrepetitive test on the prepared subgrade or on a prepared test embankment, provided that the embankment is at least 10 ft [3 m] thick. Otherwise, the test should be conducted on the subgrade, and the k-value obtained should be adjusted to account for the thickness and density of the embankment, using the nomograph provided in Step 3.

In a repetitive test, the elastic k-value is determined from the ratio of load to elastic deformation (the recoverable portion of the total deformation measured). In a nonrepetitive test, the load-deformation ratio at a deformation of 0.05 in [1.25 mm] is considered to represent the elastic k-value, according to extensive research by the U.S. Army Corps of Engineers.

Note also that a 30-in-diameter [762-mm-diameter] plate should be used to determine the elastic static k-value for use in design. Smaller diameter plates will yield substantially higher k-values, which are not appropriate for use in this design procedure.

Table 12. Determination of design subgrade k-value from deflection measurements.

BARE CONCRETE PAVEMENT

Step	Equation	Calculated Value	Example
d_0 d_8 d_{12} d_{18} d_{24} d_{36} d_{60}		——— ——— ——— ——— ——— ——— ———	0.00418 0.00398 0.00384 0.00361 0.00336 0.00288 0.00205
$AREA_7$	[26]		45.0
Initial estimate of ℓ	[28]		40.79
Nondimensional d_0* and initial estimate of k	[31] [30]		0.1237 160
AF_{d_0} AF_ℓ	[34] [35]		0.867 0.934
Adjusted k	[37]		212
Mean dynamic k			212
Mean static k for design			106

COMPOSITE PAVEMENT

Step	Equation	Calculated Value	Example
d_{12} d_{18} d_{24} d_{36} d_{60}		——— ——— ——— ——— ———	0.00349 0.00332 0.00313 0.00273 0.00202
$AREA_5$	[27]		37.8
Initial estimate of ℓ	[29]		48.83
Nondimensional d_{12}* and initial estimate of k	[33] [32]		0.1189 128
AF_{d_0} AF_ℓ	[34] [35]		0.823 0.896
Adjusted k	[37]		195
Mean dynamic k			195
Mean static k for design			97

Step 2. Determine Seasonally Adjusted Effective k-Value. The effective k-value is obtained by combining the seasonal k-values into a single "effective" value for use in concrete pavement design. The effective k-value is essentially a weighted average based on fatigue damage. The effective k-value results in the same fatigue damage over the entire year that is caused by the seasonal variation in k-value. The seasonally adjusted effective k-value is determined by the following steps:

1. Select tentative values for the slab thickness D, concrete flexural strength S'_c, concrete elastic modulus E_c, base elastic modulus E_b and friction coefficient f (both depending on base type), base thickness H_b, design temperature differential TD (for a given climatic region, as a function of the trial slab thickness D), joint spacing L, and initial and terminal serviceability P1 and P2. The tentative values selected for these parameters need only be approximate.
2. Select a k-value to represent each distinct season of the year.
3. Using each of the seasonal k-values in turn, calculate W_{18}, the allowable number of 18-kip [80-kN] ESALs for the design traffic lane, using the rigid pavement performance model given in Section 3.2.2.
4. Compute the relative damage for each season as the inverse of the calculated W_{18}.
5. Compute the total relative damage for the year and divide by the number of seasons to obtain the mean annual damage.
6. Compute a W_{18} corresponding to the mean damage as the inverse of the mean damage.
7. Use the rigid pavement performance model to determine a single k-value that produces a predicted W_{18} matching the W_{18} obtained in Step 6. This k-value is the seasonally adjusted effective k-value.

Table 13 may be used to determine the effective k-value. The example shown in Table 13 was developed using the following tentative design parameters:

D	=	9 in	[229 mm]
E_c	=	4,200,000 psi	[28,959 MPa]
S'_c	=	690 psi	[4758 kPa]
E_b	=	25,000 psi for aggregate base	[172 kPa]
H_b	=	6 in	[152 mm]
L	=	180 in	[4.57 m]
TD	=	+ 7.92°F	[+4.4°C]
P1	=	4.5	
P2	=	2.5	

Step 3. Adjust the Effective k-Value for the Effects of Embankment and/or Shallow Rigid Layer. A nomograph is provided in Figure 43 for adjustment of the seasonally adjusted effective subgrade k-value if: (a) fill material will be placed above the natural subgrade, and/or (b) a rigid layer (e.g., bedrock or hardpan clay) is present at a depth of 10 ft [3 m] or less beneath the existing subgrade surface. Note that the rigid layer adjustment should only be applied if the subgrade k was determined on the basis of soil type or similar correlations. If the k-value was determined from nondestructive deflection testing or from plate bearing tests, the effect of a rigid layer, if present at a depth of less than 10 ft [3 m], is already represented in the k-value obtained.

Table 13. Determination of seasonally adjusted effective subgrade k-value.

Season	k-value (psi/in)	W_{18} (millions)	Relative Damage $(1/W_{18})$
		Mean Damage	
		W_{18}	million
		Effective k-value	psi/in

Notes: W_{18} is computed from the rigid pavement performance model given in Section 3.2.2.
A year may be divided into as many seasons as desired to represent distinct subgrade conditions.
1 psi/in = 0.27 kPa/mm

EXAMPLE

Seasons	k-value (psi/in)	W_{18} (millions)	Relative Damage $(1/W_{18})$
Spring	100	13.18	0.0759
Summer	200	14.60	0.0685
Fall	300	15.71	0.0637
Winter	400	16.72	0.0598
		Mean Damage	0.0670
		W_{18}	14.92 million
		Effective k-value	229 psi/in

1 psi/in = 0.27 kPa/mm

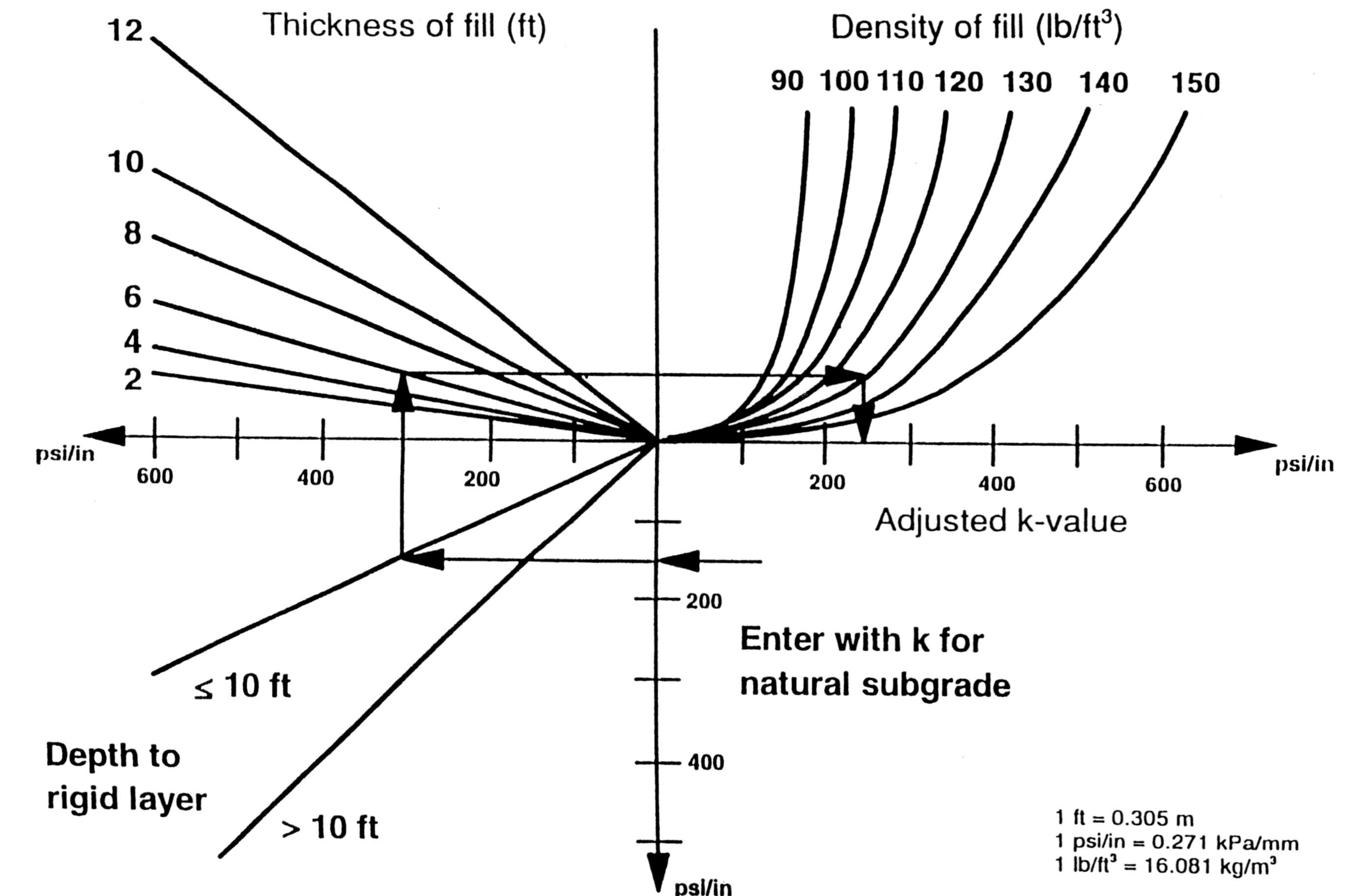

Figure 43. Adjustment to k for fill and/or rigid layer.

3.2.2 Determine Required Structural Design

A slab thickness is determined for the midslab loading position, shown in Figure 44, because for doweled pavements this is the critical fatigue damage location. Most cracks initiate at the edge of the slab as a result of this loading. This slab thickness becomes the design thickness if the transverse joints are doweled. If the joints are not doweled, a design check is made to see if the joint loading position causes a more critical stress at the top of the slab. Also, a design check is made for joint design adequacy with respect to faulting, as described in Section 3.3.

Determine Required Inputs. The following inputs must be selected or obtained.

		Section of Guide
1.	Estimated ESALs, W_{18}, for the performance period in the design lane.	2.1.2
2.	Design reliability, R, percent.	2.1.3
3.	Overall standard deviation, S_o.	2.1.3
4.	Design serviceability loss, $PSI = P_1 - P_2$.	2.2.1
5.	Effective (seasonally adjusted) elastic k-value of the subgrade, psi/in.	3.2.1
6.	Concrete modulus of rupture, S'_c, psi.	2.3.4
7.	Concrete elastic modulus, E_c, psi.	2.3.3
8.	Joint spacing, L, inches.	3.3.2
9.	Base modulus, E_b, psi.	2.3.3

10. Slab/base friction coefficient, f.

11. Base thickness, H_b, inches.

12. Effective positive temperature differential through concrete slab, TD, °F.

13. Lane edge support condition:

    a. Conventional lane width (12 ft [3.7 m]) with free edge.

    b. Conventional lane width (12 ft [3.7 m]) with tied concrete shoulder.

    c. Wide slab (e.g., 14 ft [4.3 m]) with conventional traffic lane width (12 ft [3.7 m]).

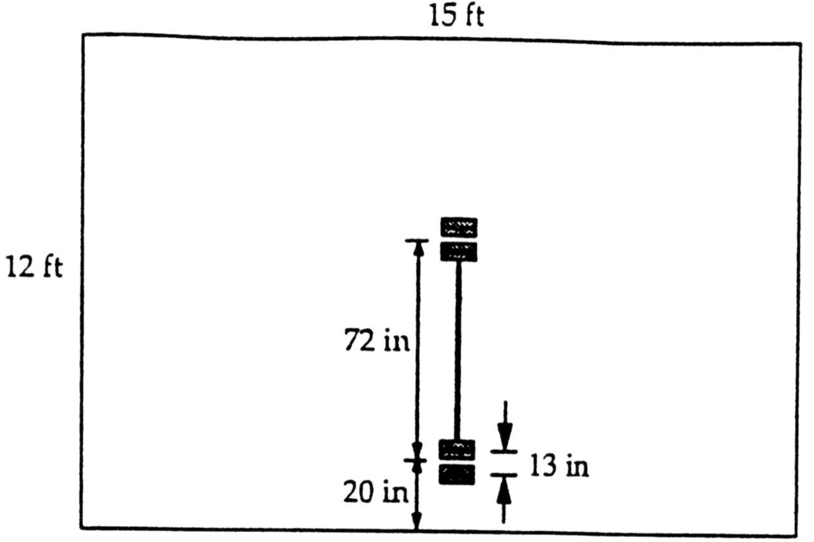

Midslab Loading

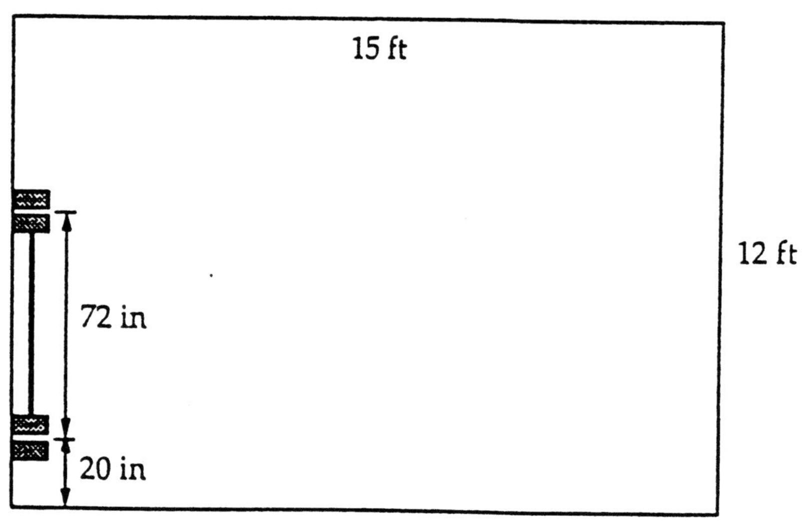

Joint Loading

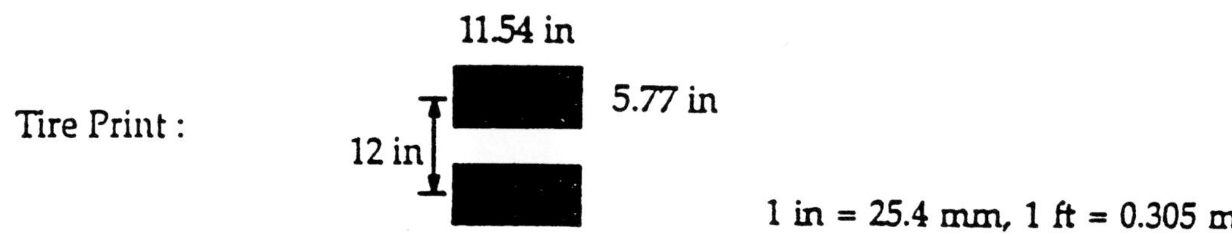

Tire Print :

1 in = 25.4 mm, 1 ft = 0.305 m

Figure 44. Midslab and joint loading positions defined.

18

Design Equations for Rigid Pavement. The rigid pavement design equation for 50 percent reliability is given below:

$$\log W' = \log W + \left(5.065 - 0.03295\, P2^{\,2.4} \right) \left[\log \left(\frac{(S'_c)'}{\sigma'_t} \right) - \log \left(\frac{690}{\sigma_t} \right) \right] \qquad [38]$$

where W$'$ = number of 18-kip [80-kN] ESALs estimated for design traffic lane
    W = number of 18-kip [80-kN] ESALs computed from Equation 39 below:
(Note: Logarithm is to base 10).

$$\log W = \log R + \frac{G}{Y} \qquad [39]$$

$$\log R = 5.85 + 7.35 \log (D + 1) - 4.62 \log (L1 + L2) + 3.28 \log L2 \qquad [40]$$

$$Y = 1.00 + \frac{3.63\,(L1 + L2)^{5.2}}{(D + 1)^{8.46}\, L2^{\,3.52}} \qquad [41]$$

$$G = \log \left(\frac{P1 - P2}{P1 - 1.5} \right) \qquad [42]$$

$\quad$ D $\;=\;$ concrete slab thickness, inches
$\quad$ L1 $\;=\;$ load on a single or tandem axle, kips
$\quad$ L2 $\;=\;$ axle code, 1 for single axle, 2 for tandem axle
$\quad$ P1 $\;=\;$ initial serviceability index
$\quad$ P2 $\;=\;$ terminal serviceability index
$(S'_c)' \;=\;$ mean 28-day, third-point loading flexural strength, psi
$\qquad\quad$ (690 psi [4758 kPa] for AASHO Road Test)
$\quad \sigma_t \;=\;$ midslab tensile stress due to load and temperature from Equation 43
$\qquad\quad$ with AASHO Road Test constants
$\quad \sigma_t' \;=\;$ midslab tensile stress due to load and temperature from Equation 43
$\qquad\quad$ with inputs for new pavement design

$$\sigma_t = \sigma_l\, E\, F \left[1.0 + 10^{(\log b)}\, TD \right] \qquad [43]$$

$\quad \sigma_l$ = midslab tensile stress due to load only, from Equation 44

$$\sigma_l = \frac{18{,}000}{D^2} \left\{ 4.227 - 2.381 \left(\frac{180}{\ell} \right)^{0.2} - 0.0015 \left[\frac{E_b \, H_b}{1.4 \, k} \right]^{0.5} - 0.155 \left[H_b \left(\frac{E_b}{E_c} \right)^{0.75} \right]^{0.5} \right\} \quad [44]$$

E_c = modulus of elasticity of concrete slab, psi
(4,200,000 psi [28,959 MPa] for AASHO Road Test)

E_b = modulus of elasticity of base, psi
(25,000 psi [172 MPa] for AASHO Road Test)

H_b = thickness of base, inches (6 in [152 mm] for AASHO Road Test)

$$\ell = \sqrt[4]{\frac{E_c \, D^3}{12 \, (1 - \mu^2) \, k}} \quad [45]$$

k = effective elastic modulus of subgrade support, psi/in
(110 psi/in [29.92 kPa/mm] for AASHO Road Test)

μ = Poisson's ratio for concrete (0.20 for AASHO Road Test)

E = edge support adjustment factor (1.00 for AASHO Road Test)
 = 1.00 for conventional 12-ft-wide [3.66-m-wide] traffic lane
 = 0.94 for conventional 12-ft-wide [3.66-m-wide] traffic lane plus tied concrete shoulder
 = 0.92 for 2-ft [0.6-m] widened slab with conventional 12-ft [3.66-m] lane width

F = ratio between slab stress at a given coefficient of friction (f)
between the slab and base and slab stress at full friction, from Equation 46

$$F = 1.177 - 4.3 * 10^{-8} \, D \, E_b - 0.01155542 \, D$$
$$+ \; 6.27 * 10^{-7} \, E_b - 0.000315 \, f \qquad [46]$$

f = friction coefficient between slab and base (see Table 14)

Table 14. Modulus of elasticity and coefficient of friction for various base types.

Base Type or Interface Treatment	Modulus of Elasticity (psi)	Peak Friction Coefficient low	mean	high
Fine-grained soil	3,000 - 40,000	0.5	1.3	2.0
Sand	10,000 - 25,000	0.5	0.8	1.0
Aggregate	15,000 - 45,000	0.7	1.4	2.0
Polyethylene sheeting	NA	0.5	0.6	1.0
Lime-stabilized clay	20,000 - 70,000	3.0	NA	5.3
Cement-treated gravel	(500 + CS) * 1000	8.0	34	63
Asphalt-treated gravel	300,000 - 600,000	3.7	5.8	10
Lean concrete without curing compound	(500 + CS) * 1000	> 36		
Lean concrete with single or double wax curing compound	(500 + CS) * 1000	3.5		4.5

Notes: CS = compressive strength, psi
Low, mean, and high measured peak coefficients of friction summarized from various references are shown above.
1 psi = 6.89 kPa

$$\log b = -1.944 + 2.279 \frac{D}{\ell} + 0.0917 \frac{L}{\ell} - 433{,}080 \frac{D^2}{k\,\ell^4}$$

$$+ \left(\frac{0.0614}{\ell} \right) * \left(\frac{E_b\,H_b^{\,1.5}}{1.4\,k} \right)^{0.5} - 438.642 \frac{D^2}{k\,\ell^2} - 498{,}240 \frac{D^3\,L}{k\,\ell^6}$$

[47]

L = joint spacing, inches (180 in [4572 mm] for AASHO Road Test)
TD = effective positive temperature differential, top of slab minus bottom of slab, °F

$$\text{effective positive } TD = 0.962 - \frac{52.181}{D} + 0.341\,WIND$$

$$+ 0.184\,TEMP - 0.00836\,PRECIP$$

[48]

D = slab thickness, inches
WIND = mean annual wind speed, mph
TEMP = mean annual temperature, °F
PRECIP = mean annual precipitation, inches

Contour maps for the three climatic inputs are provided in Figures 45, 46, and 47. In addition, these climatic data are provided for several U.S. cities in Table 15. Data for other locations are obtainable from local weather stations or other sources.

Required Slab Thickness. The rigid pavement design equations given above may be used to determine the required slab thickness for the design traffic. The design equations are too complex to put into nomograph form. However, the new design equations can easily be solved in a spreadsheet or computer program. In addition, for a given set of design inputs, a straight-line relationship exists between log W_{18} and slab thickness D:

$$D = A_0 + A_1 \log_{10} W_{18R} \qquad\qquad [49]$$

where D = required slab thickness, inches
A_0 and A_1 = regression constants dependent on other design features
W_{18R} = design 18-kip [80-kN] ESALs for the specified level of design reliability R

The W_{18R} for any level of design reliability and overall standard deviation is computed as follows:

$$W_{18R} = 10^{(\log W_{18} + Z S_o)} \qquad\qquad [50]$$

where W_{18R} = design 18-kip [80-kN] ESALs for a specified level of design reliability R
W_{18} = estimated 18-kip [80-kN] ESALs over the design period in the design lane
Z = standard deviate from normal distribution table for given level of reliability (e.g., 1.28 for R = 90 percent)
S_o = overall standard deviation

The required slab thickness D was computed for a range of joint spacings, concrete flexural strengths, subgrade k-values, and temperature differentials, for each of three base types and three levels of design reliability, as summarized below. Note that an appropriate friction coefficient for each base type was selected using Table 14.

Table	Reliability	Base Type	Base Modulus, psi
16	95	Granular	25,000
17	95	Treated	500,000
18	95	High-strength	1,000,000
19	90	Granular	25,000
20	90	Treated	500,000
21	90	High-strength	1,000,000
22	85	Granular	25,000
23	85	Treated	500,000
24	85	High-strength	1,000,000

[1 psi = 6.89 kPa]

Example designs are provided at the end of this appendix.

1 mph = 1.61 km/h

Figure 45. Mean annual wind speed, mph.

MEAN ANNUAL AIR TEMPERATURE (°F)

BASED ON NORMAL PERIOD 1961-1990

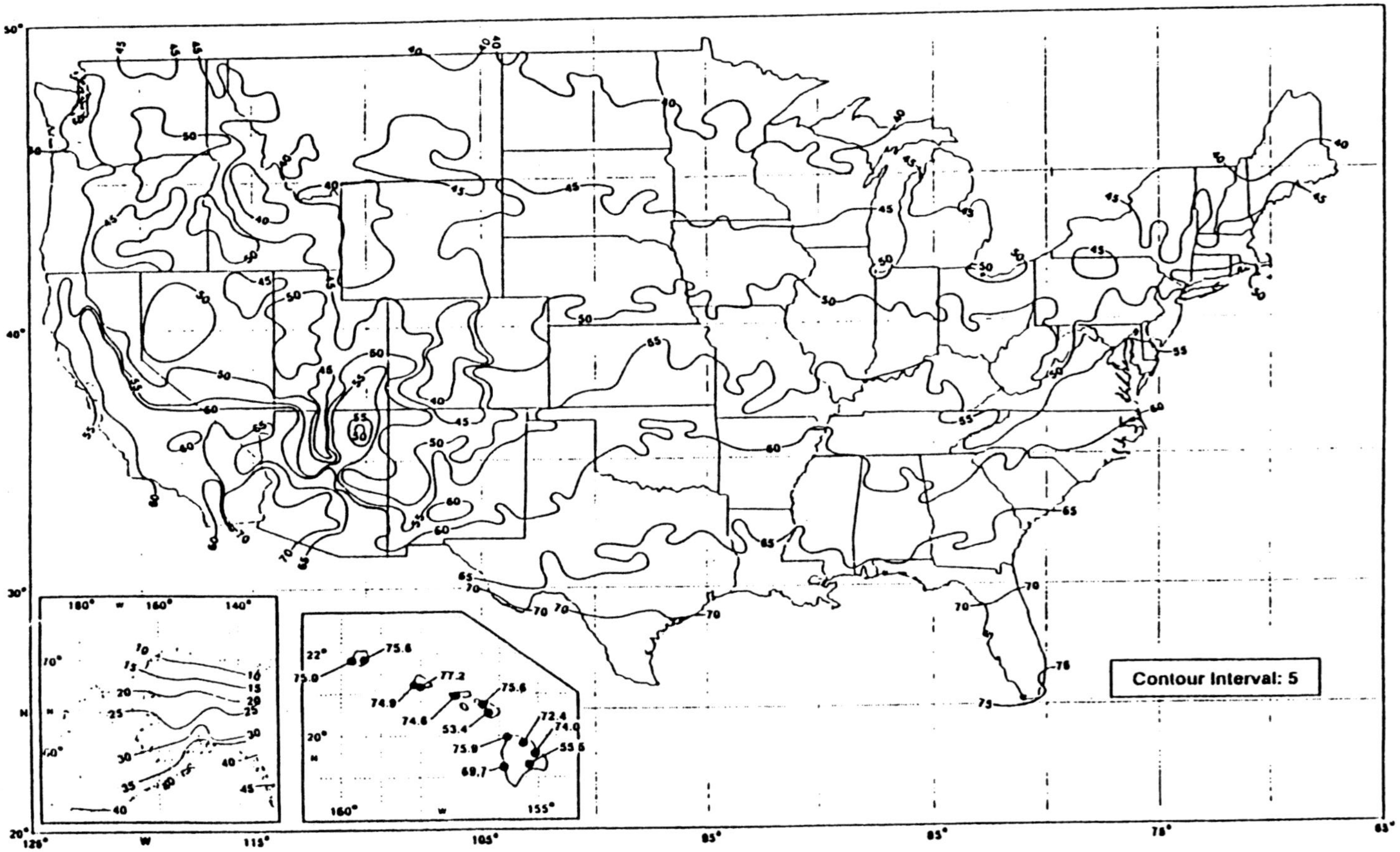

$$°C = (°F - 32)/1.8$$

Figure 46. Mean annual air temperature, °F.

MEAN ANNUAL PRECIPITATION (INCHES)

BASED ON NORMAL PERIOD 1961-1990

1 in = 25.4 mm

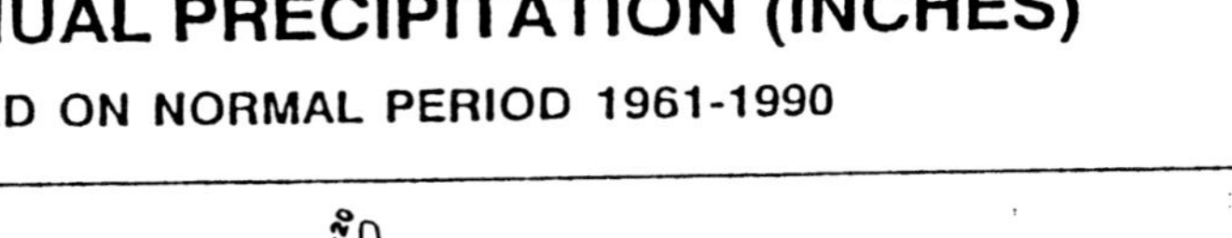

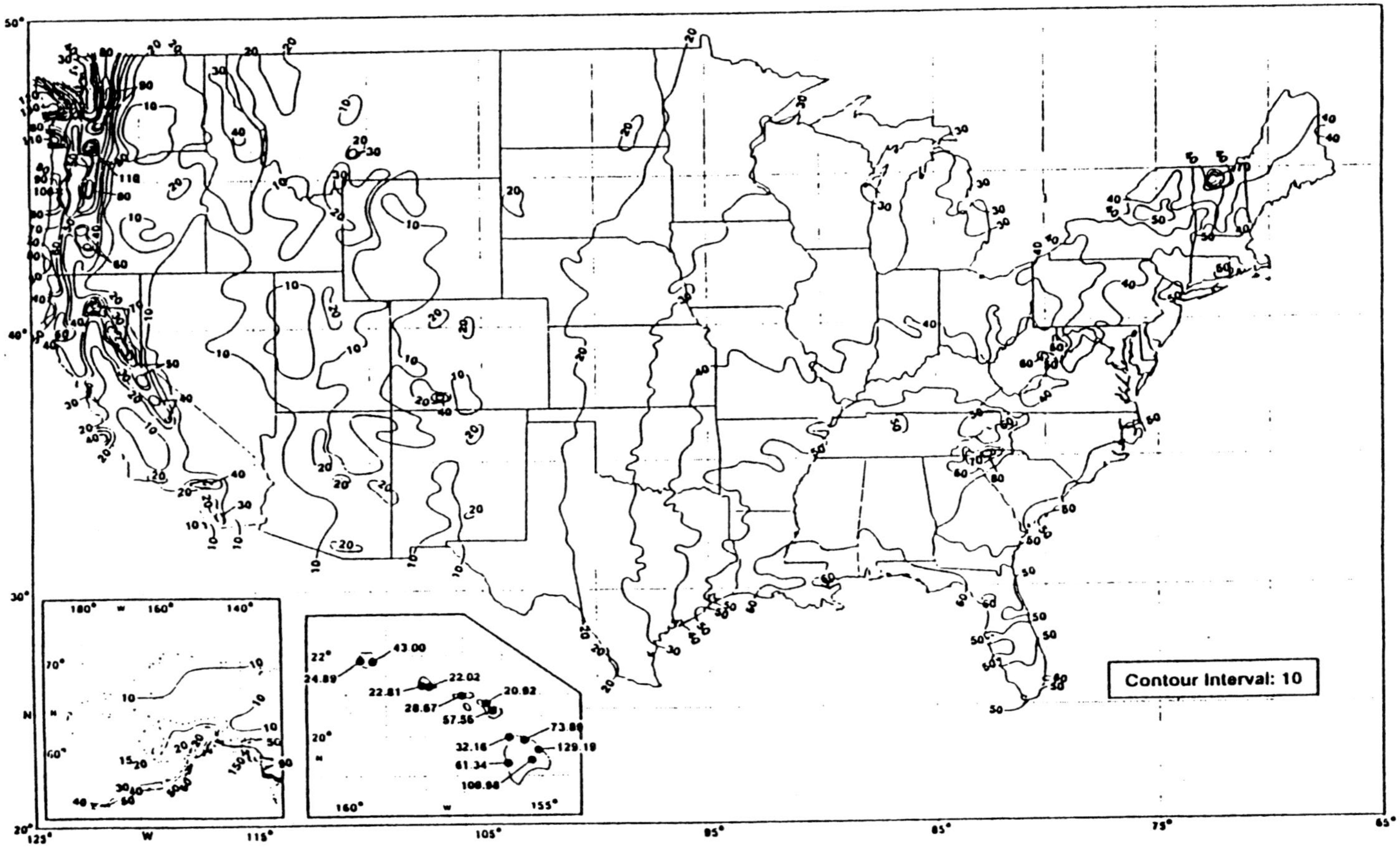

Figure 47. Mean annual precipitation, inches.

Table 15. Mean annual temperature, precipitation, and wind speed for selected U.S. cities.

Location	Mean Annual Temperature, °F	Mean Annual Precipitation, in	Mean Annual Wind Speed, mph	Location	Mean Annual Temperature, °F	Mean Annual Precipitation, in	Mean Annual Wind Speed, mph	Location	Mean Annual Temperature, °F	Mean Annual Precipitation, in	Mean Annual Wind Speed, mph
ALABAMA				KANSAS				OKLAHOMA			
Birmingham	62.2	52.2	7.2	Topeka	54.1	28.6	10.1	Oklahoma City	59.9	30.9	12.5
Mobile	67.5	64.6	9.0	Wichita	56.4	40.1	12.3	Tulsa	60.3	38.8	10.4
Montgomery	67.5	49.2	6.7	KENTUCKY				OREGON			
ALASKA				Lexington	54.9	45.7	7.1	Medford	53.6	19.8	4.8
Anchorage	35.3	15.2	6.9	Louisville	56.2	43.6	8.3	Portland	53.0	37.4	7.9
Fairbanks	25.9	10.4	5.5	LOUISIANA				Salem	52.0	40.4	7.0
King Salmon	32.8	19.3	10.8	Baton Rouge	67.5	55.8	7.7	PENNSYLVANIA			
ARIZONA				Lake Charles	68.0	53.0	8.6	Harrisburg	53.0	39.1	7.6
Flagstaff	45.4	20.9	7.1	New Orleans	68.2	59.7	8.2	Philadelphia	54.3	41.4	9.5
Phoenix	71.2	7.1	6.3	Shreveport	65.4	43.8	8.5	Pittsburgh	50.3	36.3	9.1
Tucson	68.0	11.1	8.2	MAINE				RHODE ISLAND			
ARKANSAS				Caribou	38.9	36.6	11.2	Providence	50.3	45.3	10.6
Little Rock	61.9	49.2	7.9	Portland	45.0	43.8	8.7	SOUTH CAROLINA			
CALIFORNIA				MARYLAND				Charleston	64.8	51.6	8.7
Bakersfield	65.6	5.7	6.4	Baltimore	55.1	41.8	9.2	Columbia	63.3	49.1	6.9
Fresno	62.5	10.5	6.4	MASSACHUSETTS				SOUTH DAKOTA			
Los Angeles	62.6	12.1	7.5	Boston	51.5	43.8	12.4	Huron	44.7	18.7	11.6
Sacramento	60.6	17.1	8.1	Worcester	46.8	47.6	12.4	Rapid City	46.7	16.3	11.3
San Diego	63.8	9.3	6.9	MICHIGAN				TENNESSEE			
San Francisco	56.6	19.7	10.5	Detroit	48.6	4.0	10.2	Chattanooga	59.4	52.6	6.1
Santa Barbara	58.9	16.2	6.1	Flint	46.8	29.2	10.6	Knoxville	58.9	47.3	7.1
COLORADO				Grand Rapids	47.5	34.4	9.7	Memphis	61.8	51.6	9.0
Colorado Springs	48.9	15.4	10.1	MINNESOTA				Nashville	59.2	48.5	8.0
Denver	50.3	15.3	8.8	Duluth	38.2	29.7	11.2	TEXAS			
CONNECTICUT				Minneapolis	44.7	26.4	10.6	Amarillo	57.2	19.1	13.6
Hartford	49.8	44.4	9.2	MISSISSIPPI				Brownsville	73.6	25.4	11.6
DC				Jackson	64.6	52.8	7.4	Corpus Christi	72.1	30.2	12.0
Washington	57.5	39.0	9.3	MISSOURI				Dallas	66.0	29.5	10.8
DELAWARE				Kansas City	56.3	35.2	10.7	El Paso	63.4	7.8	9.0
Wilmington	54.0	41.4	9.2	MONTANA				Galveston	69.6	40.2	11.0
FLORIDA				Great Falls	44.7	15.2	12.8	Houston	68.3	44.8	7.8
Jacksonville	68.0	52.8	8.1	NEBRASKA				Lubbock	59.9	17.8	12.4
Miami	75.6	57.6	9.2	Omaha	49.5	29.9	10.6	Midland	63.5	13.7	11.1
Orlando	72.4	47.8	8.6	NEVADA				San Antonio	68.7	29.2	9.4
Tallahassee	67.2	64.6	6.4	Las Vegas	66.3	4.2	9.2	Waco	67.0	31.0	11.3
Tampa	72.0	46.7	8.5	Reno	49.4	7.5	6.5	Wichita Falls	63.5	26.7	11.7
West Palm Beach	74.6	59.7	9.4	NEW JERSEY				UTAH			
GEORGIA				Atlantic City	53.1	41.9	10.1	Salt Lake City	51.7	15.3	8.8
Atlanta	61.2	48.6	9.1	NEW MEXICO				VERMONT			
Augusta	63.2	43.1	6.5	Albuquerque	56.2	8.1	9.0	Burlington	44.1	33.7	8.8
Macon	64.7	44.9	7.7	NEW YORK				VIRGINIA			
Savannah	65.9	49.7	7.9	Albany	47.3	35.7	8.9	Norfolk	59.5	45.2	10.6
HAWAII				Buffalo	47.6	37.5	12.1	Richmond	57.7	44.1	7.6
Hilo	73.6	128.2	7.1	New York City	54.5	44.1	12.1	Roanoke	56.1	39.2	8.2
Honolulu	77.0	23.5	11.5	Rochester	47.9	31.3	9.7	WASHINGTON			
IDAHO				Syracuse	47.7	39.1	9.7	Olympia	49.6	51.0	6.7
Boise	51.1	11.7	8.8	NORTH CAROLINA				Seattle	52.7	38.8	9.0
Pocatello	46.6	10.9	10.2	Charlotte	60.0	43.2	7.5	Spokane	47.2	16.7	8.8
ILLINOIS				Greensboro	57.9	42.5	7.5	WEST VIRGINIA			
Chicago	49.2	33.3	10.2	Raleigh	59.0	41.8	7.8	Charleston	54.8	42.4	6.4
Peoria	50.4	34.9	10.1	Wilmington	63.4	53.4	8.8	Huntington	55.2	40.7	6.5
Springfield	52.6	33.8	11.3	NORTH DAKOTA				WISCONSIN			
INDIANA				Bismarck	41.3	15.4	10.3	Green Bay	43.6	28.0	10.1
Evansville	55.7	41.6	8.2	Fargo	40.5	19.6	12.4	Madison	45.2	30.8	9.8
Fort Wayne	49.7	34.4	10.1	OHIO				Milwaukee	46.1	30.9	11.6
Indianapolis	52.1	39.1	9.6	Akron-Canton	49.5	35.9	9.8	WYOMING			
South Bend	49.4	38.2	10.4	Cleveland	49.6	35.4	10.7	Casper	45.2	11.4	13.0
IOWA				Columbus	51.7	37.0	8.7	Cheyenne	45.7	13.3	12.9
Des Moines	49.7	30.8	10.9	Dayton	51.9	34.7	10.1				
Sioux City	48.4	25.4	11.0	Youngstown	48.3	37.3	10.0				
Waterloo	46.1	33.1	10.7								

°C = (°F - 32)/1.8, 1 in = 25.4 mm, 1 mph = 1.61 km/h

Source: National Climatic Data Center, 1986

Table 16. Slab thickness computed for granular base and 95 percent reliability.

E_b = 25 ksi [172.25 MPa], R = 95 percent, S_o = 0.39, P2 = 2.5, 12-ft-wide [3.7-m wide] lanes with AC shoulders.
Computed thicknesses less than 6.0 in [152 mm] or greater than 15.0 in [381 mm] are not shown.

Joint Spacing (in)	Flexural Strength (psi)	Subgrade k (psi/in)	Positive TD (degrees F)	Design ESALs, millions							
				5	10	20	30	40	50	75	100
144	600	100	5	10.7	11.8	12.9	13.6	14.0	14.4	—	—
144	600	100	7	10.6	11.7	12.8	13.5	13.9	14.3	14.9	—
144	600	100	9	10.5	11.7	12.8	13.4	13.9	14.3	14.9	—
144	600	100	11	10.5	11.6	12.7	13.4	13.8	14.2	14.9	—
144	600	100	13	10.4	11.5	12.6	13.3	13.8	14.1	14.8	—
144	600	250	5	10.1	11.4	12.7	13.4	13.9	14.4	—	—
144	600	250	7	10.2	11.5	12.8	13.6	14.1	14.6	—	—
144	600	250	9	10.3	11.7	13.0	13.8	14.4	14.8	—	—
144	600	250	11	10.4	11.8	13.1	13.9	14.5	14.9	—	—
144	600	250	13	10.5	11.9	13.3	14.1	14.7	—	—	—
144	600	500	5	9.1	10.8	12.5	13.4	14.1	14.7	—	—
144	600	500	7	9.5	11.3	13.1	14.1	14.9	—	—	—
144	600	500	9	9.9	11.8	13.7	14.8	—	—	—	—
144	600	500	11	10.3	12.3	14.3	—	—	—	—	—
144	600	500	13	10.7	12.8	14.8	—	—	—	—	—
144	700	100	5	9.7	10.8	11.8	12.4	12.9	13.2	13.8	14.3
144	700	100	7	9.6	10.7	11.7	12.3	12.8	13.1	13.7	14.2
144	700	100	9	9.6	10.7	11.7	12.4	12.8	13.1	13.8	14.2
144	700	100	11	9.5	10.6	11.7	12.3	12.7	13.1	13.7	14.1
144	700	100	13	9.5	10.5	11.6	12.2	12.7	13.0	13.6	14.1
144	700	250	5	9.0	10.2	11.4	12.2	12.7	13.1	13.8	14.3
144	700	250	7	9.2	10.4	11.7	12.4	12.9	13.3	14.0	14.5
144	700	250	9	9.3	10.6	11.8	12.6	13.1	13.5	14.2	14.7
144	700	250	11	9.4	10.7	11.9	12.7	13.2	13.6	14.3	14.8
144	700	250	13	9.5	10.8	12.1	12.8	13.4	13.8	14.5	15.1
144	700	500	5	8.1	9.6	11.1	12.0	12.6	13.1	14.0	14.6
144	700	500	7	8.5	10.0	11.6	12.6	13.2	13.7	14.7	—
144	700	500	9	8.7	10.4	12.2	13.2	13.9	14.4	—	—
144	700	500	11	9.1	10.9	12.6	13.7	14.4	15.0	—	—
144	700	500	13	9.4	11.3	13.1	14.1	14.9	—	—	—
144	800	100	5	8.9	9.9	10.9	11.5	11.9	12.2	12.8	13.2
144	800	100	7	8.9	9.9	10.8	11.4	11.8	12.1	12.7	13.1
144	800	100	9	8.8	9.8	10.8	11.4	11.8	12.1	12.6	13.0
144	800	100	11	8.8	9.8	10.7	11.3	11.7	12.0	12.6	13.0
144	800	100	13	8.8	9.7	10.7	11.3	11.7	12.0	12.5	12.9
144	800	250	5	8.2	9.3	10.5	11.1	11.6	12.0	12.6	13.1
144	800	250	7	8.3	9.5	10.7	11.3	11.8	12.2	12.9	13.4
144	800	250	9	8.5	9.6	10.8	11.5	12.0	12.4	13.1	13.6
144	800	250	11	8.6	9.8	11.0	11.7	12.2	12.6	13.3	13.8
144	800	250	13	8.7	9.9	11.1	11.9	12.4	12.8	13.5	14.0
144	800	500	5	7.2	8.6	10.0	10.8	11.4	11.9	12.7	13.3
144	800	500	7	7.6	9.1	10.5	11.4	12.0	12.5	13.4	14.0
144	800	500	9	7.9	9.5	11.0	11.9	12.6	13.0	14.0	14.6
144	800	500	11	8.2	9.8	11.4	12.3	13.0	13.5	14.4	—
144	800	500	13	8.6	10.2	11.8	12.7	13.4	13.9	14.8	—
192	600	100	5	10.8	12.0	13.1	13.8	14.2	14.6	—	—
192	600	100	7	10.8	11.9	13.1	13.7	14.2	14.5	—	—
192	600	100	9	10.8	11.9	13.0	13.7	14.1	14.5	—	—
192	600	100	11	10.7	11.9	13.0	13.6	14.1	14.5	—	—
192	600	100	13	10.7	11.8	12.9	13.6	14.1	14.4	—	—
192	600	250	5	10.4	11.7	13.0	13.7	14.3	14.7	—	—
192	600	250	7	10.6	11.9	13.2	14.0	14.6	15.0	—	—
192	600	250	9	10.8	12.2	13.5	14.3	14.9	—	—	—
192	600	250	11	10.9	12.3	13.7	14.5	—	—	—	—
192	600	250	13	11.1	12.5	14.0	14.8	—	—	—	—
192	600	500	5	9.5	11.3	13.0	14.1	—	—	—	—
192	600	500	7	10.1	12.0	13.9	15.0	—	—	—	—
192	600	500	9	10.7	12.7	14.7	—	—	—	—	—
192	600	500	11	11.3	13.4	—	—	—	—	—	—
192	600	500	13	11.8	14.0	—	—	—	—	—	—
192	700	100	5	9.9	10.9	12.0	12.6	13.1	13.4	14.0	14.5
192	700	100	7	9.9	10.9	12.0	12.6	13.0	13.4	14.0	14.4
192	700	100	9	9.8	10.9	12.0	12.6	13.0	13.4	14.0	14.4
192	700	100	11	9.8	10.9	11.9	12.6	13.0	13.3	14.0	14.4
192	700	100	13	9.8	10.9	11.9	12.5	13.0	13.3	13.9	14.4
192	700	250	5	9.4	10.6	11.8	12.5	13.0	13.4	14.1	14.6
192	700	250	7	9.6	10.8	12.0	12.8	13.3	13.7	14.4	14.9
192	700	250	9	9.8	11.1	12.3	13.1	13.6	14.0	14.8	—
192	700	250	11	10.0	11.3	12.6	13.4	13.9	14.3	—	—
192	700	250	13	10.2	11.5	12.8	13.5	14.1	14.5	—	—

1 in = 25.4 mm, 1 psi = 6.89 kPa, 1 psi/in = .271 kPa/mm, °C = (°F-32)/1.8

Table 16. Slab thickness computed for granular base and 95 percent reliability (continued).

E_b = 25 ksi [172.25 MPa], R = 95 percent, S_o = 0.39, $P2$ = 2.5, 12-ft-wide [3.7-m wide] lanes with AC shoulders.
Computed thicknesses less than 6.0 in [152 mm] or greater than 15.0 in [381 mm] are not shown.

Joint Spacing (in)	Flexural Strength (psi)	Subgrade k (psi/in)	Positive TD (degrees F)	Design ESALs, millions							
				5	10	20	30	40	50	75	100
192	700	500	5	8.5	10.1	11.6	12.6	13.2	13.7	——	—
192	700	500	7	9.0	10.7	12.4	13.4	14.1	14.7	——	—
192	700	500	9	9.6	11.3	13.1	14.1	14.9	——	——	—
192	700	500	11	10.1	11.9	13.7	14.8	—	—	—	—
192	700	500	13	10.6	12.4	14.3	—	—	—	—	—
192	800	100	5	9.1	10.1	11.1	11.7	12.1	12.4	13.0	13.4
192	800	100	7	9.1	10.1	11.1	11.7	12.1	12.4	13.0	13.4
192	800	100	9	9.1	10.1	11.1	11.6	12.1	12.4	13.0	13.4
192	800	100	11	9.1	10.1	11.1	11.6	12.1	12.4	12.9	13.4
192	800	100	13	9.1	10.1	11.1	11.6	12.0	12.4	12.9	13.3
192	800	250	5	8.5	9.7	10.8	11.5	12.0	12.4	13.0	13.5
192	800	250	7	8.7	9.9	11.1	11.8	12.3	12.7	13.4	13.9
192	800	250	9	8.9	10.2	11.4	12.1	12.6	13.0	13.7	14.2
192	800	250	11	9.2	10.4	11.6	12.3	12.8	13.2	13.9	14.4
192	800	250	13	9.4	10.6	11.8	12.5	13.0	13.4	14.1	14.6
192	800	500	5	7.7	9.2	10.6	11.5	12.1	12.5	13.4	14.0
192	800	500	7	8.3	9.8	11.3	12.2	12.8	13.3	14.2	14.8
192	800	500	9	8.7	10.3	11.9	12.8	13.5	14.0	14.9	—
192	800	500	11	9.2	10.8	12.5	13.4	14.1	14.6	—	—
192	800	500	13	9.6	11.3	13.0	13.9	14.6	—	—	—
240	600	100	5	11.0	12.1	13.2	13.9	14.3	14.7	—	—
240	600	100	7	11.0	12.1	13.2	13.9	14.4	14.7	—	—
240	600	100	9	11.0	12.1	13.3	13.9	14.4	14.7	—	—
240	600	100	11	11.0	12.1	13.3	13.9	14.4	14.8	—	—
240	600	100	13	11.0	12.2	13.3	13.9	14.4	14.8	—	—
240	600	250	5	10.7	12.0	13.3	14.1	14.6	15.0	—	—
240	600	250	7	11.0	12.3	13.7	14.5	15.0	—	—	—
240	600	250	9	11.3	12.7	14.1	14.9	—	—	—	—
240	600	250	11	11.6	13.0	14.5	—	—	—	—	—
240	600	250	13	11.8	13.2	14.7	—	—	—	—	—
240	600	500	5	10.0	11.9	13.7	—	—	—	—	—
240	600	500	7	10.8	12.8	14.8	—	—	—	—	—
240	600	500	9	11.7	13.8	—	—	—	—	—	—
240	600	500	11	12.4	14.6	—	—	—	—	—	—
240	600	500	13	13.2	—	—	—	—	—	—	—
240	700	100	5	10.1	11.1	12.2	12.8	13.3	13.6	14.2	14.7
240	700	100	7	10.1	11.1	12.2	12.8	13.2	13.6	14.2	14.6
240	700	100	9	10.1	11.2	12.2	12.8	13.3	13.6	14.2	14.7
240	700	100	11	10.1	11.2	12.2	12.9	13.3	13.6	14.2	14.7
240	700	100	13	10.2	11.2	12.3	12.9	13.3	13.7	14.3	14.7
240	700	250	5	9.7	10.9	12.1	12.9	13.4	13.8	14.5	15.0
240	700	250	7	10.0	11.3	12.6	13.3	13.8	14.3	15.0	—
240	700	250	9	10.3	11.6	12.9	13.7	14.2	14.6	—	—
240	700	250	11	10.6	11.9	13.2	14.0	14.5	14.9	—	—
240	700	250	13	10.9	12.2	13.5	14.3	14.9	—	—	—
240	700	500	5	9.0	10.7	12.3	13.3	14.0	—	—	—
240	700	500	7	9.8	11.6	13.3	14.3	15.0	—	—	—
240	700	500	9	10.6	12.4	14.2	—	—	—	—	—
240	700	500	11	11.2	13.1	14.9	—	—	—	—	—
240	700	500	13	11.9	13.7	—	—	—	—	—	—
240	800	100	5	9.3	10.3	11.3	11.9	12.3	12.6	13.2	13.6
240	800	100	7	9.3	10.3	11.3	11.9	12.3	12.7	13.2	13.7
240	800	100	9	9.4	10.4	11.4	12.0	12.4	12.7	13.3	13.7
240	800	100	11	9.4	10.4	11.4	12.0	12.4	12.8	13.4	13.8
240	800	100	13	9.5	10.5	11.5	12.1	12.5	12.8	13.4	13.8
240	800	250	5	8.8	10.0	11.2	11.9	12.4	12.8	13.5	14.0
240	800	250	7	9.3	10.4	11.6	12.3	12.8	13.2	13.8	14.3
240	800	250	9	9.6	10.8	12.0	12.7	13.2	13.6	14.3	14.8
240	800	250	11	9.8	11.0	12.2	13.0	13.5	13.8	14.5	15.0
240	800	250	13	10.1	11.3	12.5	13.3	13.8	14.2	14.9	15.4
240	800	500	5	8.3	9.8	11.3	12.1	12.7	13.2	14.1	14.7
240	800	500	7	9.0	10.6	12.2	13.1	13.7	14.2	—	—
240	800	500	9	9.7	11.3	12.9	13.9	14.5	15.0	—	—
240	800	500	11	10.4	12.0	13.6	14.6	—	—	—	—
240	800	500	13	10.9	12.5	14.2	—	—	—	—	—

1 in = 25.4 mm, 1 psi = 6.89 kPa, 1 psi/in = .271 kPa/mm, °C = (°F-32)/1.8

Table 17. Slab thickness computed for treated base and 95 percent reliability.

E_b = 500 ksi [3445 MPa], R = 95 percent, S_o = 0.39, P2 = 2.5, 12-ft-wide [3.7-m wide] lanes with AC shoulders.
Computed thicknesses less than 6.0 in [152 mm] or greater than 15.0 in [381 mm] are not shown.

Joint Spacing (in)	Flexural Strength (psi)	Subgrade k (psi/in)	Positive TD (degrees F)	Design ESALs, millions							
				5	10	20	30	40	50	75	100
144	600	100	5	9.2	10.3	11.3	12.0	12.4	12.7	13.4	13.8
144	600	100	7	9.4	10.4	11.4	12.1	12.5	12.8	13.4	13.9
144	600	100	9	9.5	10.5	11.5	12.1	12.6	12.9	13.5	13.9
144	600	100	11	9.6	10.6	11.6	12.2	12.6	13.0	13.6	14.0
144	600	100	13	9.6	10.7	11.7	12.3	12.7	13.0	13.6	14.1
144	600	250	5	8.6	10.0	11.3	12.1	12.6	13.0	13.8	14.3
144	600	250	7	9.1	10.4	11.7	12.5	13.0	13.4	14.2	14.7
144	600	250	9	9.4	10.7	12.1	12.8	13.4	13.8	14.6	—
144	600	250	11	9.7	11.0	12.4	13.1	13.7	14.1	14.9	—
144	600	250	13	10.0	11.3	12.6	13.4	14.0	14.4	—	—
144	600	500	5	7.5	9.2	10.9	11.9	12.6	13.1	—	—
144	600	500	7	8.4	10.1	11.8	12.8	13.5	14.0	—	—
144	600	500	9	9.1	10.8	12.5	13.5	14.2	14.8	—	—
144	600	500	11	9.4	11.3	13.2	14.3	—	—	—	—
144	600	500	13	10.1	11.9	13.8	14.9	—	—	—	—
144	700	100	5	8.5	9.5	10.4	11.0	11.4	11.8	12.3	12.7
144	700	100	7	8.6	9.6	10.6	11.1	11.5	11.9	12.4	12.8
144	700	100	9	8.8	9.7	10.7	11.2	11.6	12.0	12.5	12.9
144	700	100	11	8.9	9.8	10.8	11.3	11.7	12.0	12.6	13.0
144	700	100	13	9.0	9.9	10.9	11.4	11.8	12.1	12.7	13.0
144	700	250	5	8.0	9.2	10.3	11.0	11.5	11.9	12.6	13.1
144	700	250	7	8.3	9.5	10.7	11.5	12.0	12.4	13.1	13.6
144	700	250	9	8.7	9.9	11.1	11.8	12.3	12.7	13.4	13.9
144	700	250	11	9.0	10.2	11.4	12.1	12.6	13.0	13.7	14.2
144	700	250	13	9.2	10.4	11.7	12.4	12.9	13.3	14.0	14.5
144	700	500	5	6.6	8.2	9.8	10.8	11.4	12.0	12.9	13.6
144	700	500	7	7.6	9.1	10.7	11.6	12.3	12.8	13.7	14.3
144	700	500	9	8.3	9.8	11.4	12.3	12.9	13.4	14.4	15.0
144	700	500	11	8.6	10.3	12.0	12.9	13.6	14.2	—	—
144	700	500	13	9.2	10.9	12.5	13.4	14.1	14.6	—	—
144	800	100	5	7.9	8.8	9.7	10.3	10.7	11.0	11.5	11.9
144	800	100	7	8.0	8.9	9.9	10.4	10.8	11.1	11.6	12.0
144	800	100	9	8.2	9.1	10.0	10.5	10.9	11.2	11.7	12.1
144	800	100	11	8.3	9.2	10.1	10.6	11.0	11.3	11.8	12.1
144	800	100	13	8.4	9.3	10.2	10.7	11.1	11.3	11.8	12.2
144	800	250	5	7.3	8.5	9.6	10.2	10.7	11.1	11.7	12.2
144	800	250	7	7.7	8.9	10.0	10.6	11.1	11.4	12.1	12.6
144	800	250	9	8.1	9.2	10.3	11.0	11.4	11.8	12.5	12.9
144	800	250	11	8.4	9.5	10.6	11.3	11.7	12.1	12.7	13.2
144	800	250	13	8.6	9.7	10.9	11.5	12.0	12.3	13.0	13.5
144	800	500	5	6.4	7.7	9.1	9.9	10.5	11.0	11.8	12.4
144	800	500	7	6.9	8.4	9.8	10.7	11.3	11.7	12.6	13.2
144	800	500	9	7.6	9.1	10.5	11.3	11.9	12.4	13.2	13.8
144	800	500	11	7.9	9.5	11.0	11.9	12.5	13.0	13.9	14.5
144	800	500	13	8.5	10.0	11.5	12.4	13.0	13.5	14.3	14.9
192	600	100	5	9.5	10.5	11.6	12.2	12.6	13.0	13.6	14.0
192	600	100	7	9.7	10.7	11.7	12.3	12.8	13.1	13.7	14.2
192	600	100	9	9.8	10.9	11.9	12.5	12.9	13.3	13.9	14.3
192	600	100	11	9.9	11.0	12.0	12.6	13.0	13.3	13.9	14.3
192	600	100	13	10.1	11.1	12.1	12.7	13.1	13.4	14.0	14.4
192	600	250	5	9.1	10.4	11.7	12.5	13.0	13.4	14.2	14.7
192	600	250	7	9.6	10.9	12.2	13.0	13.5	13.9	14.7	—
192	600	250	9	10.0	11.3	12.6	13.4	14.0	14.4	15.1	—
192	600	250	11	10.4	11.7	13.0	13.8	14.3	14.8	15.5	—
192	600	250	13	10.6	12.0	13.5	14.3	15.0	15.7	16.3	—
192	600	500	5	8.1	9.8	11.5	12.5	13.2	13.8	14.8	—
192	600	500	7	9.1	10.9	12.6	13.6	14.3	14.9	—	—
192	600	500	9	9.8	11.7	13.5	14.6	—	—	—	—
192	600	500	11	10.6	12.5	14.3	—	—	—	—	—
192	600	500	13	11.3	13.2	—	—	—	—	—	—
192	700	100	5	8.7	9.7	10.7	11.3	11.7	12.0	12.5	12.9
192	700	100	7	9.0	9.9	10.9	11.4	11.8	12.2	12.7	13.1
192	700	100	9	9.1	10.1	11.0	11.6	12.0	12.3	12.9	13.2
192	700	100	11	9.3	10.2	11.2	11.7	12.1	12.4	13.0	13.4
192	700	100	13	9.4	10.4	11.3	11.9	12.2	12.5	13.1	13.5
192	700	250	5	8.3	9.5	10.8	11.5	12.0	12.4	13.1	13.6
192	700	250	7	8.9	10.1	11.3	12.0	12.5	12.9	13.6	14.1
192	700	250	9	9.3	10.5	11.7	12.4	12.9	13.3	14.0	14.5
192	700	250	11	9.7	10.9	12.1	12.8	13.3	13.7	14.4	14.9
192	700	250	13	9.9	11.2	12.4	13.1	13.6	14.0	14.8	—

1 in = 25.4 mm, 1 psi = 6.89 kPa, 1 psi/in = .271 kPa/mm, °C = (°F-32)/1.8

Table 17. Slab thickness computed for treated base and 95 percent reliability (continued).

E_b = 500 ksi [3445 MPa], R = 95 percent, S_o = 0.39, P2 = 2.5, 12-ft-wide [3.7-m wide] lanes with AC shoulders.

Computed thicknesses less than 6.0 in [152 mm] or greater than 15.0 in [381 mm] are not shown.

Joint Spacing (in)	Flexural Strength (psi)	Subgrade k (psi/in)	Positive TD (degrees F)	Design ESALs, millions							
				5	10	20	30	40	50	75	100
192	700	500	5	7.5	9.0	10.6	11.5	12.1	12.6	13.5	—
192	700	500	7	8.5	10.0	11.5	12.4	13.1	13.6	14.5	—
192	700	500	9	9.1	10.7	12.3	13.3	14.0	14.5	—	—
192	700	500	11	9.9	11.5	13.1	14.0	14.7	—	—	—
192	700	500	13	10.5	12.1	13.7	14.6	—	—	—	—
192	800	100	5	8.1	9.0	10.0	10.5	10.9	11.2	11.7	12.1
192	800	100	7	8.4	9.3	10.2	10.7	11.1	11.4	11.9	12.3
192	800	100	9	8.6	9.5	10.4	10.9	11.2	11.5	12.0	12.4
192	800	100	11	8.8	9.6	10.5	11.0	11.4	11.7	12.2	12.5
192	800	100	13	8.9	9.8	10.6	11.1	11.5	11.8	12.3	12.6
192	800	250	5	7.8	8.9	10.0	10.7	11.1	11.5	12.1	12.6
192	800	250	7	8.3	9.4	10.5	11.2	11.6	12.0	12.6	13.1
192	800	250	9	8.7	9.8	10.9	11.6	12.1	12.4	13.1	13.5
192	800	250	11	9.1	10.2	11.3	12.0	12.4	12.8	13.4	13.9
192	800	250	13	9.4	10.5	11.6	12.3	12.7	13.1	13.8	14.2
192	800	500	5	6.9	8.3	9.8	10.6	11.2	11.7	12.5	13.1
192	800	500	7	7.9	9.3	10.7	11.5	12.1	12.6	13.4	14.0
192	800	500	9	8.5	10.0	11.4	12.3	12.9	13.4	14.2	14.8
192	800	500	11	9.2	10.7	12.1	13.0	13.6	14.0	14.9	—
192	800	500	13	9.6	11.1	12.7	13.6	14.2	14.7	—	—
240	600	100	5	9.7	10.8	11.8	12.4	12.9	13.2	13.8	14.2
240	600	100	7	10.0	11.0	12.0	12.6	13.1	13.4	14.0	14.4
240	600	100	9	10.2	11.2	12.2	12.8	13.3	13.6	14.2	14.6
240	600	100	11	10.4	11.4	12.4	13.0	13.4	13.7	14.3	14.8
240	600	100	13	10.5	11.5	12.5	13.1	13.5	13.8	14.4	14.8
240	600	250	5	9.5	10.8	12.1	12.9	13.4	13.9	14.6	—
240	600	250	7	10.1	11.4	12.8	13.5	14.1	14.5	—	—
240	600	250	9	10.7	12.0	13.3	14.0	14.6	15.0	—	—
240	600	250	11	11.0	12.4	13.8	14.6	—	—	—	—
240	600	250	13	11.5	12.8	14.2	15.0	—	—	—	—
240	600	500	5	8.8	10.6	12.3	13.3	14.0	—	—	—
240	600	500	7	9.8	11.7	13.6	14.7	—	—	—	—
240	600	500	9	10.9	12.8	14.6	—	—	—	—	—
240	600	500	11	11.8	13.7	—	—	—	—	—	—
240	600	500	13	12.5	14.5	—	—	—	—	—	—
240	700	100	5	9.0	10.0	11.0	11.5	11.9	12.2	12.8	13.2
240	700	100	7	9.3	10.3	11.2	11.8	12.2	12.5	13.0	13.4
240	700	100	9	9.5	10.5	11.4	12.0	12.4	12.7	13.2	13.6
240	700	100	11	9.7	10.7	11.6	12.2	12.6	12.9	13.4	13.8
240	700	100	13	9.9	10.8	11.8	12.3	12.7	13.0	13.6	14.0
240	700	250	5	8.8	10.0	11.2	11.9	12.4	12.8	13.5	14.0
240	700	250	7	9.5	10.7	11.8	12.5	13.0	13.4	14.1	14.6
240	700	250	9	9.9	11.1	12.4	13.1	13.6	14.0	14.7	—
240	700	250	11	10.4	11.6	12.8	13.5	14.0	14.4	—	—
240	700	250	13	10.8	12.0	13.2	13.9	14.4	14.7	—	—
240	700	500	5	8.3	9.8	11.3	12.2	12.9	13.4	14.3	14.9
240	700	500	7	9.2	10.8	12.5	13.4	14.1	14.6	—	—
240	700	500	9	10.3	11.8	13.4	14.3	15.0	—	—	—
240	700	500	11	11.0	12.7	14.4	—	—	—	—	—
240	700	500	13	11.7	13.4	—	—	—	—	—	—
240	800	100	5	8.5	9.4	10.3	10.8	11.1	11.4	12.0	12.3
240	800	100	7	8.8	9.6	10.5	11.0	11.4	11.7	12.2	12.6
240	800	100	9	9.0	9.9	10.8	11.3	11.6	11.9	12.4	12.8
240	800	100	11	9.2	10.1	10.9	11.5	11.8	12.1	12.6	13.0
240	800	100	13	9.4	10.2	11.1	11.6	12.0	12.3	12.8	13.1
240	800	250	5	8.3	9.4	10.5	11.1	11.6	12.0	12.6	13.1
240	800	250	7	8.9	10.0	11.1	11.8	12.2	12.6	13.2	13.7
240	800	250	9	9.4	10.5	11.6	12.3	12.7	13.1	13.7	14.2
240	800	250	11	9.8	10.9	12.0	12.7	13.1	13.5	14.1	14.6
240	800	250	13	10.2	11.3	12.4	13.0	13.5	13.8	14.4	14.9
240	800	500	5	7.8	9.2	10.6	11.4	12.0	12.4	13.2	13.8
240	800	500	7	8.8	10.2	11.6	12.5	13.1	13.5	14.4	14.9
240	800	500	9	9.7	11.1	12.5	13.3	13.9	14.4	—	—
240	800	500	11	10.4	11.8	13.2	14.1	14.7	—	—	—
240	800	500	13	11.0	12.4	13.9	14.7	—	—	—	—

1 in = 25.4 mm, 1 psi = 6.89 kPa, 1 psi/in = .271 kPa/mm, °C = (°F-32)/1.8

Table 18. Slab thickness computed for high-strength base and 95 percent reliability.

E_b = 1 million psi [6890 MPa], R = 95 percent, S_o = 0.39, P2 = 2.5, 12-ft-wide [3.7-m wide] lanes with AC shoulders.
Computed thicknesses less than 6.0 in [152 mm] or greater than 15.0 in [381 mm] are not shown.

Joint Spacing (in)	Flexural Strength (psi)	Subgrade k (psi/in)	Positive TD (degrees F)	Design ESALs, millions							
				5	10	20	30	40	50	75	100
144	600	100	5	8.5	9.5	10.5	11.2	11.6	11.9	12.5	13.0
144	600	100	7	8.8	9.8	10.8	11.4	11.8	12.1	12.7	13.1
144	600	100	9	9.0	10.0	11.0	11.6	12.0	12.3	12.9	13.3
144	600	100	11	9.2	10.2	11.1	11.7	12.1	12.5	13.0	13.4
144	600	100	13	9.4	10.3	11.3	11.9	12.3	12.6	13.1	13.5
144	600	250	5	8.1	9.4	10.7	11.4	12.0	12.4	13.1	13.7
144	600	250	7	8.8	10.0	11.2	12.0	12.5	12.9	13.6	14.1
144	600	250	9	9.3	10.5	11.7	12.4	12.9	13.3	14.0	14.5
144	600	250	11	9.6	10.8	12.0	12.8	13.3	13.7	14.4	14.9
144	600	250	13	9.9	11.2	12.4	13.1	13.6	14.0	14.7	15.2
144	600	500	5	7.4	8.9	10.4	11.3	11.9	12.4	13.3	13.9
144	600	500	7	7.9	9.6	11.2	12.2	12.9	13.5	14.4	—
144	600	500	9	8.9	10.5	12.1	13.0	13.7	14.2	—	—
144	600	500	11	9.7	11.2	12.8	13.6	14.3	14.8	—	—
144	600	500	13	10.0	11.7	13.4	14.4	—	—	—	—
144	700	100	5	7.9	8.9	9.8	10.4	10.8	11.1	11.7	12.1
144	700	100	7	8.2	9.2	10.1	10.7	11.0	11.3	11.9	12.3
144	700	100	9	8.5	9.4	10.3	10.8	11.2	11.5	12.1	12.4
144	700	100	11	8.7	9.6	10.5	11.0	11.4	11.7	12.2	12.6
144	700	100	13	8.8	9.7	10.6	11.2	11.5	11.8	12.3	12.7
144	700	250	5	7.5	8.7	9.9	10.6	11.1	11.5	12.2	12.8
144	700	250	7	8.2	9.3	10.5	11.2	11.6	12.0	12.7	13.1
144	700	250	9	8.4	9.6	10.8	11.6	12.1	12.5	13.2	13.7
144	700	250	11	9.0	10.1	11.2	11.9	12.4	12.8	13.4	13.9
144	700	250	13	9.3	10.4	11.6	12.2	12.7	13.0	13.7	14.2
144	700	500	5	6.8	8.2	9.6	10.5	11.1	11.5	12.4	12.9
144	700	500	7	7.3	8.9	10.4	11.3	11.9	12.4	13.3	13.9
144	700	500	9	8.4	9.8	11.2	12.0	12.6	13.1	13.9	14.5
144	700	500	11	9.1	10.5	11.8	12.6	13.2	13.7	14.5	15.0
144	700	500	13	9.4	10.9	12.3	13.2	13.8	14.3	—	—
144	800	100	5	7.4	8.3	9.3	9.8	10.2	10.5	11.0	11.4
144	800	100	7	7.7	8.6	9.5	10.0	10.4	10.7	11.2	11.6
144	800	100	9	8.0	8.9	9.7	10.2	10.6	10.9	11.4	11.7
144	800	100	11	8.2	9.1	9.9	10.4	10.8	11.0	11.5	11.9
144	800	100	13	8.4	9.2	10.1	10.6	10.9	11.2	11.7	12.0
144	800	250	5	6.9	8.1	9.3	10.0	10.5	10.8	11.5	12.0
144	800	250	7	7.7	8.8	9.8	10.5	10.9	11.3	11.9	12.3
144	800	250	9	8.0	9.1	10.2	10.9	11.3	11.7	12.4	12.8
144	800	250	11	8.5	9.5	10.6	11.2	11.7	12.0	12.6	13.1
144	800	250	13	8.8	9.9	10.9	11.5	11.9	12.3	12.9	13.3
144	800	500	5	6.2	7.6	8.9	9.7	10.3	10.7	11.5	12.1
144	800	500	7	6.8	8.2	9.7	10.5	11.1	11.6	12.5	13.1
144	800	500	9	7.9	9.2	10.5	11.3	11.8	12.2	13.0	13.5
144	800	500	11	8.5	9.8	11.1	11.8	12.4	12.8	13.5	14.1
144	800	500	13	8.9	10.2	11.5	12.3	12.8	13.3	14.0	14.6
192	600	100	5	8.8	9.8	10.8	11.4	11.8	12.2	12.8	13.2
192	600	100	7	9.2	10.1	11.1	11.7	12.1	12.4	13.0	13.4
192	600	100	9	9.4	10.4	11.4	11.9	12.3	12.7	13.2	13.6
192	600	100	11	9.6	10.6	11.6	12.1	12.5	12.8	13.4	13.8
192	600	100	13	9.9	10.8	11.7	12.3	12.7	13.0	13.5	13.9
192	600	250	5	8.7	9.9	11.1	11.9	12.4	12.8	13.5	14.0
192	600	250	7	9.4	10.6	11.8	12.5	13.0	13.3	14.0	14.5
192	600	250	9	9.8	11.0	12.3	13.0	13.5	13.9	14.6	—
192	600	250	11	10.3	11.5	12.7	13.4	13.9	14.3	15.0	—
192	600	250	13	10.7	11.9	13.1	13.8	14.3	14.6	—	—
192	600	500	5	8.2	9.6	11.1	12.0	12.6	13.0	13.9	—
192	600	500	7	8.9	10.5	12.1	13.0	13.7	14.2	—	—
192	600	500	9	10.0	11.5	13.0	13.9	14.5	15.0	—	—
192	600	500	11	10.5	12.2	13.8	14.7	—	—	—	—
192	600	500	13	11.3	12.9	14.4	—	—	—	—	—
192	700	100	5	8.2	9.2	10.1	10.7	11.1	11.4	11.9	12.3
192	700	100	7	8.6	9.5	10.4	11.0	11.4	11.7	12.2	12.6
192	700	100	9	8.9	9.8	10.7	11.2	11.6	11.9	12.4	12.8
192	700	100	11	9.2	10.0	10.9	11.4	11.8	12.1	12.6	13.0
192	700	100	13	9.4	10.2	11.1	11.6	12.0	12.2	12.3	13.1
192	700	250	5	8.1	9.3	10.4	11.1	11.6	11.9	12.6	13.1
192	700	250	7	8.6	9.8	11.0	11.7	12.2	12.6	13.2	13.7
192	700	250	9	9.3	10.4	11.5	12.2	12.6	13.0	13.6	14.1
192	700	250	11	9.8	10.8	11.9	12.6	13.0	13.4	14.0	14.5
192	700	250	13	10.1	11.2	12.3	12.9	13.4	13.7	14.4	14.8

31

1 in = 25.4 mm, 1 psi = 6.89 kPa, 1 psi/in = .271 kPa/mm, °C = (°F-32)/1.8

Table 18. Slab thickness computed for high-strength base and 95 percent reliability (continued).

E_b = 1 million psi [6890 MPa], R = 95 percent, S_o = 0.39, P2 = 2.5, 12-ft-wide [3.7-m wide] lanes with AC shoulders.
Computed thicknesses less than 6.0 in [152 mm] or greater than 15.0 in [381 mm] are not shown.

Joint Spacing (in)	Flexural Strength (psi)	Subgrade k (psi/in)	Positive TD (degrees F)	Design ESALs, millions							
				5	10	20	30	40	50	75	100
192	700	500	5	7.6	9.0	10.4	11.2	11.7	12.2	13.0	13.5
192	700	500	7	8.6	9.9	11.3	12.1	12.7	13.2	14.0	14.5
192	700	500	9	9.1	10.6	12.1	13.0	13.7	14.1	15.0	—
192	700	500	11	10.0	11.4	12.8	13.6	14.2	14.7	—	—
192	700	500	13	10.7	12.0	13.4	14.2	14.8	—	—	—
192	800	100	5	7.8	8.7	9.6	10.1	10.5	10.7	11.3	11.6
192	800	100	7	8.2	9.0	9.9	10.4	10.7	11.0	11.5	11.9
192	800	100	9	8.5	9.3	10.2	10.6	11.0	11.3	11.8	12.1
192	800	100	11	8.7	9.5	10.4	10.9	11.2	11.5	11.9	12.3
192	800	100	13	8.9	9.7	10.6	11.0	11.4	11.6	12.1	12.4
192	800	250	5	7.7	8.7	9.8	10.4	10.9	11.2	11.9	12.3
192	800	250	7	8.2	9.3	10.4	11.0	11.5	11.8	12.5	12.9
192	800	250	9	8.8	9.9	10.9	11.5	11.9	12.3	12.9	13.3
192	800	250	11	9.3	10.3	11.3	11.9	12.3	12.6	13.2	13.7
192	800	250	13	9.6	10.6	11.6	12.2	12.7	13.0	13.6	14.0
192	800	500	5	7.2	8.5	9.7	10.5	11.0	11.4	12.2	12.7
192	800	500	7	8.1	9.4	10.7	11.4	12.0	12.4	13.1	13.7
192	800	500	9	8.8	10.1	11.4	12.2	12.7	13.1	13.9	14.5
192	800	500	11	9.5	10.8	12.1	12.8	13.4	13.8	14.5	15.0
192	800	500	13	9.9	11.3	12.6	13.4	14.0	14.4	—	—
240	600	100	5	9.1	10.1	11.1	11.7	12.1	12.4	13.0	13.4
240	600	100	7	9.5	10.5	11.5	12.0	12.4	12.8	13.3	13.8
240	600	100	9	9.8	10.8	11.8	12.3	12.7	13.0	13.6	14.0
240	600	100	11	10.1	11.1	12.0	12.5	12.9	13.2	13.8	14.2
240	600	100	13	10.4	11.3	12.2	12.7	13.1	13.4	14.0	14.3
240	600	250	5	9.2	10.4	11.6	12.3	12.8	13.2	13.9	14.4
240	600	250	7	9.9	11.1	12.3	13.0	13.5	13.9	14.7	—
240	600	250	9	10.5	11.7	12.9	13.6	14.1	14.5	—	—
240	600	250	11	11.0	12.2	13.4	14.1	14.5	14.9	—	—
240	600	250	13	11.4	12.6	13.9	14.6	—	—	—	—
240	600	500	5	8.4	10.1	11.7	12.6	13.3	13.8	14.8	—
240	600	500	7	9.9	11.5	13.0	13.9	14.5	—	—	—
240	600	500	9	10.8	12.4	14.1	—	—	—	—	—
240	600	500	11	11.7	13.4	15.0	—	—	—	—	—
240	600	500	13	12.4	14.2	—	—	—	—	—	—
240	700	100	5	8.6	9.5	10.4	11.0	11.4	11.7	12.2	12.6
240	700	100	7	9.0	9.9	10.8	11.3	11.7	12.0	12.5	12.9
240	700	100	9	9.4	10.2	11.1	11.6	12.0	12.3	12.8	13.2
240	700	100	11	9.6	10.5	11.4	11.9	12.2	12.5	13.0	13.4
240	700	100	13	9.9	10.7	11.6	12.1	12.4	12.7	13.2	13.6
240	700	250	5	8.7	9.8	10.9	11.6	12.0	12.4	13.0	13.5
240	700	250	7	9.4	10.5	11.6	12.3	12.7	13.1	13.7	14.2
240	700	250	9	10.0	11.1	12.2	12.8	13.3	13.6	14.2	14.7
240	700	250	11	10.4	11.5	12.7	13.3	13.8	14.1	14.8	—
240	700	250	13	10.9	12.0	13.0	13.7	14.1	14.5	15.1	—
240	700	500	5	8.2	9.6	11.0	11.9	12.4	12.9	13.7	14.3
240	700	500	7	9.2	10.7	12.2	13.1	13.7	14.2	—	—
240	700	500	9	10.4	11.8	13.1	13.9	14.5	14.9	—	—
240	700	500	11	11.0	12.5	13.9	14.8	—	—	—	—
240	700	500	13	11.8	13.2	14.6	—	—	—	—	—
240	800	100	5	8.2	9.0	9.9	10.4	10.5	11.0	11.5	11.9
240	800	100	7	8.6	9.4	10.3	10.8	11.1	11.4	11.9	12.2
240	800	100	9	8.9	9.8	10.6	11.1	11.4	11.7	12.1	12.5
240	800	100	11	9.2	10.0	10.8	11.3	11.6	11.9	12.4	12.7
240	800	100	13	9.5	10.3	11.0	11.5	11.8	12.1	12.6	12.9
240	800	250	5	8.1	9.2	10.3	11.0	11.4	11.8	12.4	12.9
240	800	250	7	9.0	10.0	11.0	11.6	12.0	12.4	13.0	13.4
240	800	250	9	9.5	10.5	11.6	12.2	12.6	12.9	13.5	14.0
240	800	250	11	10.0	11.0	12.0	12.6	13.0	13.3	13.9	14.3
240	800	250	13	10.4	11.4	12.4	13.0	13.4	13.7	14.3	14.7
240	800	500	5	7.9	9.2	10.5	11.2	11.8	12.2	12.9	13.5
240	800	500	7	9.0	10.3	11.5	12.3	12.8	13.2	14.0	14.5
240	800	500	9	10.0	11.2	12.4	13.2	13.7	14.1	14.8	—
240	800	500	11	10.6	11.9	13.1	13.9	14.4	14.8	—	—
240	800	500	13	11.2	12.5	13.7	14.4	14.9	—	—	—

1 in = 25.4 mm, 1 psi = 6.89 kPa, 1 psi/in = .271 kPa/mm, °C = (°F-32)/1.8

Table 19. Slab thickness computed for granular base and 90 percent reliability.

E_b = 25 ksi [172.25 MPa], R = 90 percent, S_o = 0.39, P2 = 2.5, 12-ft-wide [3.7-m wide] lanes with AC shoulders.
Computed thicknesses less than 6.0 in [152 mm] or greater than 15.0 in [381 mm] are not shown.

Joint Spacing (in)	Flexural Strength (psi)	Subgrade k (psi/in)	Positive TD (degrees F)	Design ESALs, millions							
				1.5	2	2.5	3	4	5	7.5	10
144	600	100	5	8.2	8.7	9.0	9.3	9.8	10.2	10.8	11.3
144	600	100	7	8.1	8.6	9.0	9.2	9.7	10.1	10.7	11.2
144	600	100	9	8.0	8.5	8.9	9.2	9.6	10.0	10.7	11.1
144	600	100	11	8.0	8.4	8.8	9.1	9.6	9.9	10.6	11.0
144	600	100	13	7.9	8.4	8.7	9.0	9.5	9.9	10.5	11.0
144	600	250	5	7.3	7.8	8.2	8.5	9.1	9.5	10.2	10.8
144	600	250	7	7.3	7.9	8.3	8.6	9.2	9.6	10.4	10.9
144	600	250	9	7.4	7.9	8.4	8.7	9.3	9.7	10.5	11.1
144	600	250	11	7.4	8.0	8.4	8.8	9.4	9.8	10.6	11.2
144	600	250	13	7.5	8.1	8.5	8.9	9.4	9.9	10.7	11.3
144	600	500	5	—	6.1	6.7	7.1	7.8	8.3	9.3	10.0
144	600	500	7	—	6.3	6.9	7.4	8.1	8.7	9.7	10.5
144	600	500	9	—	6.5	7.1	7.6	8.4	9.0	10.1	10.9
144	600	500	11	6.0	6.8	7.4	7.9	8.8	9.4	10.5	11.4
144	600	500	13	6.1	7.0	7.7	8.2	9.1	9.7	10.9	11.8
144	700	100	5	7.4	7.8	8.2	8.4	8.9	9.2	9.8	10.3
144	700	100	7	7.3	7.8	8.1	8.4	8.8	9.1	9.8	10.2
144	700	100	9	7.2	7.7	8.0	8.3	8.7	9.1	9.7	10.2
144	700	100	11	7.2	7.6	8.0	8.3	8.7	9.0	9.7	10.1
144	700	100	13	7.1	7.6	7.9	8.2	8.6	9.0	9.6	10.0
144	700	250	5	6.3	6.8	7.2	7.5	8.0	8.4	9.1	9.6
144	700	250	7	6.5	7.0	7.4	7.7	8.3	8.6	9.4	9.9
144	700	250	9	6.6	7.1	7.5	7.8	8.4	8.8	9.5	10.0
144	700	250	11	6.7	7.2	7.6	7.9	8.4	8.8	9.6	10.1
144	700	250	13	6.7	7.3	7.7	8.0	8.5	8.9	9.7	10.2
144	700	500	5	—	—	—	6.2	6.9	7.3	8.2	8.9
144	700	500	7	—	—	6.1	6.5	7.2	7.7	8.6	9.3
144	700	500	9	—	—	6.2	6.7	7.4	7.9	8.9	9.6
144	700	500	11	—	—	6.5	7.0	7.7	8.3	9.3	10.0
144	700	500	13	—	6.2	6.8	7.2	8.0	8.6	9.6	10.4
144	800	100	5	6.7	7.2	7.5	7.7	8.1	8.5	9.0	9.4
144	800	100	7	6.7	7.1	7.4	7.7	8.1	8.4	9.0	9.4
144	800	100	9	6.7	7.1	7.4	7.6	8.1	8.4	8.9	9.3
144	800	100	11	6.6	7.0	7.4	7.6	8.0	8.3	8.9	9.3
144	800	100	13	6.6	7.0	7.3	7.6	8.0	8.3	8.9	9.3
144	800	250	5	—	6.1	6.5	6.8	7.3	7.7	8.3	8.8
144	800	250	7	—	6.2	6.6	6.9	7.4	7.8	8.5	8.9
144	800	250	9	—	6.3	6.7	7.0	7.5	7.9	8.6	9.1
144	800	250	11	—	6.4	6.8	7.1	7.6	8.0	8.7	9.2
144	800	250	13	—	6.5	6.9	7.2	7.7	8.1	8.8	9.3
144	800	500	5	—	—	—	—	6.1	6.5	7.4	7.9
144	800	500	7	—	—	—	—	6.4	6.9	7.7	8.4
144	800	500	9	—	—	—	6.0	6.7	7.2	8.1	8.7
144	800	500	11	—	—	—	6.3	7.0	7.5	8.4	9.1
144	800	500	13	—	—	6.2	6.6	7.3	7.8	8.7	9.4
192	600	100	5	8.4	8.8	9.2	9.5	10.0	10.3	11.0	11.4
192	600	100	7	8.3	8.8	9.1	9.4	9.9	10.3	10.9	11.4
192	600	100	9	8.3	8.7	9.1	9.4	9.9	10.2	10.9	11.4
192	600	100	11	8.2	8.7	9.1	9.4	9.8	10.2	10.9	11.3
192	600	100	13	8.2	8.7	9.0	9.3	9.8	10.2	10.8	11.3
192	600	250	5	7.5	8.0	8.4	8.8	9.3	9.7	10.5	11.0
192	600	250	7	7.6	8.2	8.6	9.0	9.5	9.9	10.7	11.3
192	600	250	9	7.7	8.3	8.8	9.1	9.7	10.1	10.9	11.5
192	600	250	11	7.9	8.5	8.9	9.3	9.8	10.3	11.1	11.7
192	600	250	13	8.0	8.6	9.0	9.4	10.0	10.5	11.3	11.9
192	600	500	5	—	6.4	6.9	7.4	8.1	8.7	9.7	10.4
192	600	500	7	—	6.7	7.3	7.8	8.6	9.2	10.3	11.1
192	600	500	9	6.3	7.1	7.8	8.3	9.1	9.8	10.9	11.8
192	600	500	11	6.6	7.5	8.2	8.7	9.6	10.3	11.5	12.4
192	600	500	13	7.0	7.9	8.6	9.2	10.1	10.8	12.1	13.0
192	700	100	5	7.5	8.0	8.3	8.6	9.0	9.4	10.0	10.4
192	700	100	7	7.5	8.0	8.3	8.6	9.0	9.4	10.0	10.4
192	700	100	9	7.5	7.9	8.3	8.6	9.0	9.3	10.0	10.4
192	700	100	11	7.5	7.9	8.3	8.5	9.0	9.3	9.9	10.4
192	700	100	13	7.5	7.9	8.3	8.5	9.0	9.3	9.9	10.4
192	700	250	5	6.7	7.2	7.6	7.9	8.4	8.8	9.5	10.0
192	700	250	7	6.9	7.4	7.8	8.1	8.6	9.0	9.7	10.2
192	700	250	9	7.0	7.5	7.9	8.3	8.8	9.2	9.9	10.5
192	700	250	11	7.1	7.7	8.1	8.4	9.0	9.4	10.1	10.7
192	700	250	13	7.3	7.8	8.2	8.6	9.1	9.5	10.3	10.9

1 in = 25.4 mm, 1 psi = 6.89 kPa, 1 psi/in = .271 kPa/mm, °C = (°F-32)/1.8

Table 19. Slab thickness computed for granular base and 90 percent reliability (continued).

E_b = 25 ksi [172.25 MPa], R = 90 percent, S_o = 0.39, P2 = 2.5, 12-ft-wide [3.7-m wide] lanes with AC shoulders.

Computed thicknesses less than 6.0 in [152 mm] or greater than 15.0 in [381 mm] are not shown.

Joint Spacing (in)	Flexural Strength (psi)	Subgrade k (psi/in)	Positive TD (degrees F)	Design ESALs, millions							
				1.5	2	2.5	3	4	5	7.5	10
192	700	500	5	—	—	6.2	6.6	7.3	7.8	8.7	9.3
192	700	500	7	—	—	6.5	6.9	7.7	8.2	9.2	9.9
192	700	500	9	—	6.4	7.0	7.4	8.2	8.7	9.8	10.5
192	700	500	11	6.1	6.8	7.4	7.9	8.6	9.2	10.3	11.1
192	700	500	13	6.5	7.2	7.8	8.3	9.1	9.7	10.8	11.6
192	800	100	5	6.9	7.3	7.6	7.9	8.3	8.6	9.2	9.6
192	800	100	7	6.9	7.3	7.6	7.9	8.3	8.6	9.2	9.6
192	800	100	9	6.9	7.3	7.6	7.9	8.3	8.6	9.2	9.6
192	800	100	11	6.9	7.3	7.6	7.9	8.3	8.6	9.2	9.6
192	800	100	13	6.9	7.3	7.7	7.9	8.3	8.6	9.2	9.6
192	800	250	5	—	6.4	6.8	7.1	7.6	7.9	8.6	9.1
192	800	250	7	6.1	6.6	7.0	7.3	7.8	8.2	8.9	9.4
192	800	250	9	6.2	6.7	7.1	7.5	8.0	8.4	9.1	9.6
192	800	250	11	6.6	7.1	7.5	7.8	8.3	8.6	9.3	9.8
192	800	250	13	6.7	7.2	7.6	7.9	8.4	8.8	9.5	10.0
192	800	500	5	—	—	—	—	6.5	7.0	7.9	8.5
192	800	500	7	—	—	6.0	6.4	7.1	7.5	8.4	9.1
192	800	500	9	—	—	6.4	6.8	7.4	8.0	8.9	9.5
192	800	500	11	—	6.3	6.8	7.2	7.9	8.4	9.4	10.1
192	800	500	13	6.0	6.7	7.2	7.6	8.3	8.9	9.8	10.5
240	600	100	5	8.5	9.0	9.4	9.6	10.1	10.5	11.1	11.6
240	600	100	7	8.5	9.0	9.4	9.7	10.1	10.5	11.1	11.6
240	600	100	9	8.5	9.0	9.4	9.7	10.1	10.5	11.1	11.6
240	600	100	11	8.5	9.0	9.4	9.7	10.1	10.5	11.2	11.6
240	600	100	13	8.5	9.0	9.4	9.7	10.1	10.5	11.2	11.6
240	600	250	5	7.7	8.3	8.7	9.1	9.6	10.0	10.8	11.4
240	600	250	7	8.0	8.5	9.0	9.3	9.9	10.3	11.1	11.7
240	600	250	9	8.2	8.8	9.2	9.6	10.2	10.6	11.4	12.0
240	600	250	11	8.4	9.0	9.4	9.8	10.4	10.9	11.7	12.3
240	600	250	13	8.6	9.2	9.6	10.0	10.6	11.1	11.9	12.5
240	600	500	5	—	6.7	7.3	7.8	8.6	9.2	10.2	11.0
240	600	500	7	6.5	7.3	7.9	8.4	9.3	9.9	11.1	11.9
240	600	500	9	7.0	7.9	8.6	9.1	10.0	10.7	11.9	12.8
240	600	500	11	7.6	8.5	9.2	9.8	10.7	11.4	12.7	13.6
240	600	500	13	8.1	9.1	9.8	10.4	11.4	12.1	13.4	14.4
240	700	100	5	7.7	8.1	8.5	8.8	9.2	9.5	10.2	10.6
240	700	100	7	7.8	8.2	8.5	8.8	9.2	9.6	10.2	10.6
240	700	100	9	7.8	8.2	8.6	8.9	9.3	9.6	10.2	10.7
240	700	100	11	7.8	8.3	8.6	8.9	9.3	9.7	10.3	10.7
240	700	100	13	7.9	8.3	8.6	8.9	9.3	9.7	10.3	10.7
240	700	250	5	7.0	7.5	7.9	8.2	8.7	9.1	9.8	10.3
240	700	250	7	7.2	7.8	8.2	8.5	9.0	9.4	10.2	10.7
240	700	250	9	7.5	8.0	8.4	8.8	9.3	9.7	10.5	11.0
240	700	250	11	7.7	8.3	8.7	9.0	9.6	10.0	10.8	11.3
240	700	250	13	7.9	8.5	8.9	9.3	9.8	10.2	11.0	11.6
240	700	500	5	—	6.0	6.5	7.0	7.7	8.2	9.2	9.9
240	700	500	7	6.0	6.7	7.2	7.7	8.4	9.0	10.0	10.7
240	700	500	9	6.6	7.3	7.9	8.4	9.1	9.7	10.8	11.5
240	700	500	11	7.2	7.9	8.5	9.0	9.8	10.4	11.5	12.2
240	700	500	13	7.7	8.5	9.1	9.6	10.4	11.0	12.1	12.9
240	800	100	5	7.1	7.5	7.8	8.1	8.5	8.8	9.4	9.8
240	800	100	7	7.1	7.5	7.9	8.1	8.5	8.9	9.5	9.9
240	800	100	9	7.2	7.6	7.9	8.2	8.6	8.9	9.5	9.9
240	800	100	11	7.2	7.7	8.0	8.2	8.7	9.0	9.6	10.0
240	800	100	13	7.3	7.7	8.0	8.3	8.7	9.0	9.6	10.0
240	800	250	5	6.2	6.7	7.1	7.4	7.9	8.3	9.0	9.5
240	800	250	7	6.7	7.2	7.6	7.9	8.3	8.7	9.4	9.9
240	800	250	9	6.9	7.4	7.8	8.1	8.6	9.0	9.7	10.2
240	800	250	11	7.2	7.7	8.1	8.4	8.9	9.3	10.0	10.5
240	800	250	13	7.4	7.9	8.3	8.6	9.1	9.5	10.2	10.7
240	800	500	5	—	—	6.1	6.5	7.1	7.6	8.5	9.1
240	800	500	7	—	6.2	6.7	7.1	7.8	8.3	9.2	9.8
240	800	500	9	6.2	6.9	7.4	7.8	8.5	9.0	9.9	10.6
240	800	500	11	6.8	7.4	8.0	8.4	9.1	9.6	10.5	11.2
240	800	500	13	7.3	7.9	8.5	8.9	9.6	10.1	11.1	11.8

1 in = 25.4 mm, 1 psi = 6.89 kPa, 1 psi/in = .271 kPa/mm, °C = (°F-32)/1.8

Table 20. Slab thickness computed for treated base and 90 percent reliability.

E_b = 500 ksi [3445 MPa], R = 90 percent, S_o = 0.39, P2 = 2.5, 12-ft-wide [3.7-m wide] lanes with AC shoulders.
Computed thicknesses less than 6.0 in [152 mm] or greater than 15.0 in [381 mm] are not shown.

Joint Spacing (in)	Flexural Strength (psi)	Subgrade k (psi/in)	Positive TD (degrees F)	Design ESALs, millions							
				1.5	2	2.5	3	4	5	7.5	10
144	600	100	5	6.9	7.4	7.7	8.0	8.4	8.7	9.4	9.8
144	600	100	7	7.1	7.5	7.8	8.1	8.5	8.9	9.5	9.9
144	600	100	9	7.2	7.6	8.0	8.2	8.7	9.0	9.6	10.0
144	600	100	11	7.3	7.7	8.1	8.3	8.8	9.1	9.7	10.1
144	600	100	13	7.4	7.8	8.1	8.4	8.8	9.2	9.8	10.2
144	600	250	5	—	6.3	6.7	7.0	7.6	8.0	8.8	9.3
144	600	250	7	6.2	6.7	7.1	7.5	8.0	8.5	9.2	9.8
144	600	250	9	6.5	7.1	7.5	7.8	8.4	8.8	9.6	10.1
144	600	250	11	6.8	7.4	7.8	8.1	8.7	9.1	9.9	10.4
144	600	250	13	7.1	7.6	8.1	8.4	9.0	9.4	10.2	10.7
144	600	500	5	—	—	—	—	6.2	6.7	7.7	8.4
144	600	500	7	—	—	—	6.3	7.0	7.5	8.5	9.3
144	600	500	9	—	6.0	6.6	7.0	7.7	8.3	9.3	10.0
144	600	500	11	—	6.0	6.6	7.1	7.9	8.5	9.7	10.4
144	600	500	13	6.0	6.8	7.4	7.9	8.6	9.2	10.3	11.1
144	700	100	5	6.3	6.7	7.0	7.3	7.7	8.0	8.6	9.0
144	700	100	7	6.5	6.9	7.2	7.5	7.9	8.2	8.8	9.2
144	700	100	9	6.7	7.1	7.4	7.6	8.0	8.3	8.9	9.3
144	700	100	11	6.8	7.2	7.5	7.7	8.1	8.4	9.0	9.4
144	700	100	13	6.9	7.3	7.6	7.8	8.2	8.5	9.1	9.5
144	700	250	5	—	—	6.2	6.5	7.0	7.4	8.1	8.6
144	700	250	7	—	6.1	6.5	6.8	7.3	7.7	8.4	8.9
144	700	250	9	6.0	6.5	6.9	7.2	7.7	8.1	8.8	9.3
144	700	250	11	6.3	6.8	7.2	7.5	8.0	8.4	9.1	9.6
144	700	250	13	6.5	7.0	7.4	7.8	8.3	8.7	9.4	9.9
144	700	500	5	—	—	—	—	—	—	6.8	7.5
144	700	500	7	—	—	—	5.7	6.4	6.9	7.8	8.4
144	700	500	9	—	—	6.0	6.4	7.0	7.5	8.4	9.1
144	700	500	11	—	—	6.1	6.6	7.3	7.8	8.8	9.5
144	700	500	13	—	6.3	6.8	7.2	7.9	8.4	9.4	10.1
144	800	100	5	—	6.2	6.5	6.7	7.1	7.4	8.0	8.4
144	800	100	7	6.0	6.4	6.7	6.9	7.3	7.6	8.1	8.5
144	800	100	9	6.2	6.5	6.8	7.1	7.4	7.7	8.3	8.6
144	800	100	11	6.4	6.8	7.0	7.3	7.6	7.9	8.4	8.8
144	800	100	13	6.5	6.9	7.2	7.4	7.7	8.0	8.5	8.9
144	800	250	5	—	—	—	6.0	6.4	6.8	7.5	7.9
144	800	250	7	—	—	6.1	6.4	6.8	7.2	7.9	8.3
144	800	250	9	—	6.0	6.4	6.7	7.2	7.5	8.2	8.7
144	800	250	11	—	6.3	6.7	7.0	7.5	7.8	8.5	8.9
144	800	250	13	6.1	6.6	7.0	7.3	7.7	8.1	8.7	9.2
144	800	500	5	—	—	—	—	—	—	6.5	7.1
144	800	500	7	—	—	—	—	—	6.2	7.1	7.7
144	800	500	9	—	—	—	—	6.5	6.9	7.8	8.4
144	800	500	11	—	—	—	6.1	6.7	7.2	8.1	8.7
144	800	500	13	—	—	6.3	6.7	7.3	7.8	8.7	9.3
192	600	100	5	7.1	7.6	7.9	8.2	8.6	9.0	9.6	10.0
192	600	100	7	7.4	7.8	8.1	8.4	8.8	9.2	9.8	10.2
192	600	100	9	7.5	8.0	8.3	8.6	9.0	9.3	9.9	10.4
192	600	100	11	7.7	8.1	8.5	8.7	9.1	9.5	10.1	10.5
192	600	100	13	7.8	8.3	8.6	8.8	9.3	9.6	10.2	10.6
192	600	250	5	6.2	6.7	7.1	7.5	8.0	8.4	9.2	9.7
192	600	250	7	6.7	7.2	7.6	8.0	8.5	9.0	9.7	10.3
192	600	250	9	7.1	7.7	8.1	8.4	9.0	9.4	10.2	10.7
192	600	250	11	7.5	8.0	8.5	8.8	9.4	9.8	10.5	11.1
192	600	250	13	7.4	8.0	8.4	8.8	9.4	9.9	10.7	11.4
192	600	500	5	—	—	—	6.1	6.8	7.3	8.3	9.0
192	600	500	7	—	6.1	6.6	7.1	7.8	8.3	9.3	10.1
192	600	500	9	—	6.4	7.0	7.5	8.3	8.9	10.0	10.8
192	600	500	11	6.5	7.3	7.9	8.4	9.1	9.7	10.8	11.6
192	600	500	13	7.1	7.9	8.5	9.0	9.8	10.4	11.5	12.3
192	700	100	5	6.6	7.0	7.3	7.6	8.0	8.3	8.9	9.3
192	700	100	7	6.8	7.2	7.5	7.8	8.2	8.5	9.1	9.5
192	700	100	9	7.0	7.4	7.7	8.0	8.4	8.7	9.2	9.6
192	700	100	11	7.2	7.6	7.9	8.2	8.5	8.9	9.4	9.8
192	700	100	13	7.4	7.7	8.0	8.3	8.7	9.0	9.5	9.9
192	700	250	5	—	6.1	6.5	6.8	7.4	7.7	8.5	9.0
192	700	250	7	6.2	6.7	7.1	7.4	7.9	8.3	9.0	9.5
192	700	250	9	6.7	7.2	7.5	7.9	8.4	8.7	9.4	9.9
192	700	250	11	7.0	7.5	7.9	8.2	8.7	9.1	9.8	10.3
192	700	250	13	7.2	7.7	8.1	8.4	8.9	9.3	10.1	10.6

1 in = 25.4 mm, 1 psi = 6.89 kPa, 1 psi/in = .271 kPa/mm, °C = (°F-32)/1.8

Table 20. Slab thickness computed for treated base and 90 percent reliability (continued).

E_b = 500 ksi [3445 MPa], R = 90 percent, S_o = 0.39, P2 = 2.5, 12-ft-wide [3.7-m wide] lanes with AC shoulders.

Computed thicknesses less than 6.0 in [152 mm] or greater than 15.0 in [381 mm] are not shown.

Joint Spacing (in)	Flexural Strength (psi)	Subgrade k (psi/in)	Positive TD (degrees F)	Design ESALs, millions							
				1.5	2	2.5	3	4	5	7.5	10
192	700	500	5	—	—	—	—	6.3	6.8	7.7	8.3
192	700	500	7	—	—	6.2	6.6	7.3	7.7	8.6	9.3
192	700	500	9	—	6.2	6.7	7.1	7.8	8.3	9.3	10.0
192	700	500	11	6.3	7.0	7.5	7.9	8.6	9.1	10.0	10.7
192	700	500	13	6.9	7.6	8.1	8.5	9.2	9.7	10.7	11.3
192	800	100	5	6.1	6.5	6.8	7.0	7.4	7.7	8.2	8.6
192	800	100	7	6.4	6.8	7.1	7.3	7.7	8.0	8.5	8.9
192	800	100	9	6.6	7.0	7.3	7.5	7.9	8.2	8.7	9.1
192	800	100	11	6.8	7.2	7.5	7.7	8.1	8.3	8.9	9.2
192	800	100	13	7.0	7.3	7.6	7.8	8.2	8.5	9.0	9.3
192	800	250	5	—	—	6.2	6.5	6.9	7.3	7.9	8.4
192	800	250	7	—	6.3	6.6	6.9	7.4	7.8	8.4	8.9
192	800	250	9	6.3	6.7	7.1	7.4	7.8	8.2	8.8	9.3
192	800	250	11	6.6	7.1	7.4	7.7	8.2	8.6	9.2	9.7
192	800	250	13	6.9	7.3	7.7	8.0	8.5	8.8	9.5	10.0
192	800	500	5	—	—	—	—	—	6.2	7.1	7.7
192	800	500	7	—	—	—	6.2	6.8	7.2	8.1	8.6
192	800	500	9	—	—	6.4	6.8	7.4	7.8	8.7	9.3
192	800	500	11	6.1	6.6	7.1	7.5	8.1	8.6	9.4	10.0
192	800	500	13	6.2	6.9	7.4	7.8	8.4	8.9	9.8	10.4
240	600	100	5	7.4	7.8	8.2	8.5	8.9	9.2	9.8	10.3
240	600	100	7	7.7	8.1	8.5	8.7	9.2	9.5	10.1	10.5
240	600	100	9	7.9	8.4	8.7	9.0	9.4	9.7	10.3	10.7
240	600	100	11	8.1	8.5	8.9	9.1	9.6	9.9	10.5	10.9
240	600	100	13	8.3	8.7	9.0	9.3	9.7	10.0	10.6	11.0
240	600	250	5	6.6	7.2	7.6	7.9	8.5	8.9	9.7	10.2
240	600	250	7	7.2	7.8	8.2	8.6	9.1	9.5	10.3	10.8
240	600	250	9	7.8	8.3	8.7	9.1	9.6	10.0	10.8	11.4
240	600	250	11	8.0	8.5	9.0	9.4	9.9	10.4	11.2	11.8
240	600	250	13	8.4	9.0	9.5	9.8	10.4	10.8	11.6	12.2
240	600	500	5	—	—	6.3	6.8	7.5	8.0	9.0	9.8
240	600	500	7	—	6.4	7.0	7.5	8.3	8.9	10.0	10.8
240	600	500	9	6.9	7.6	8.2	8.7	9.5	10.1	11.1	11.9
240	600	500	11	7.7	8.5	9.1	9.6	10.4	11.0	12.0	12.8
240	600	500	13	8.1	8.9	9.6	10.1	10.9	11.6	12.8	13.6
240	700	100	5	6.9	7.3	7.6	7.9	8.3	8.6	9.1	9.5
240	700	100	7	7.2	7.6	7.9	8.2	8.5	8.9	9.4	9.8
240	700	100	9	7.4	7.8	8.1	8.4	8.8	9.1	9.6	10.0
240	700	100	11	7.7	8.0	8.3	8.6	9.0	9.3	9.8	10.2
240	700	100	13	7.8	8.2	8.5	8.8	9.2	9.5	10.0	10.4
240	700	250	5	6.2	6.7	7.1	7.4	7.9	8.3	9.0	9.5
240	700	250	7	6.8	7.3	7.7	8.0	8.5	8.9	9.6	10.1
240	700	250	9	7.2	7.7	8.1	8.4	8.9	9.3	10.0	10.5
240	700	250	11	7.7	8.2	8.6	8.9	9.4	9.8	10.5	11.0
240	700	250	13	8.1	8.6	9.0	9.3	9.8	10.2	10.9	11.4
240	700	500	5	—	—	6.0	6.4	7.1	7.6	8.5	9.1
240	700	500	7	—	6.4	6.9	7.3	8.0	8.5	9.4	10.1
240	700	500	9	6.8	7.4	8.0	8.4	9.0	9.5	10.4	11.1
240	700	500	11	7.2	7.9	8.5	8.9	9.6	10.2	11.2	11.9
240	700	500	13	8.0	8.7	9.3	9.7	10.4	10.9	11.9	12.6
240	800	100	5	6.5	6.8	7.1	7.4	7.7	8.0	8.6	8.9
240	800	100	7	6.8	7.2	7.5	7.7	8.1	8.3	8.9	9.2
240	800	100	9	7.1	7.4	7.7	7.9	8.3	8.6	9.1	9.5
240	800	100	11	7.3	7.6	7.9	8.1	8.5	8.8	9.3	9.7
240	800	100	13	7.5	7.8	8.1	8.3	8.7	9.0	9.5	9.8
240	800	250	5	—	6.3	6.6	6.9	7.4	7.8	8.4	8.9
240	800	250	7	6.5	6.9	7.3	7.6	8.0	8.4	9.0	9.5
240	800	250	9	6.9	7.4	7.7	8.0	8.5	8.9	9.5	10.0
240	800	250	11	7.4	7.9	8.2	8.5	9.0	9.3	10.0	10.4
240	800	250	13	7.8	8.2	8.6	8.9	9.3	9.7	10.3	10.8
240	800	500	5	—	—	—	6.1	6.7	7.2	8.0	8.5
240	800	500	7	—	6.2	6.7	7.1	7.6	8.1	8.9	9.5
240	800	500	9	6.6	7.2	7.6	8.0	8.6	9.0	9.9	10.4
240	800	500	11	7.2	7.8	8.2	8.6	9.2	9.7	10.5	11.1
240	800	500	13	7.9	8.5	8.9	9.3	9.9	10.3	11.2	11.8

1 in = 25.4 mm, 1 psi = 6.89 kPa, 1 psi/in = .271 kPa/mm, °C = (°F-32)/1.8

Table 21. Slab thickness computed for high-strength base and 90 percent reliability.

E_b = 1 million psi [6890 MPa], R = 90 percent, S_o = 0.39, P2 = 2.5, 12-ft-wide [3.7-m wide] lanes with AC shoulders.
Computed thicknesses less than 6.0 in [152 mm] or greater than 15.0 in [381 mm] are not shown.

Joint Spacing (in)	Flexural Strength (psi)	Subgrade k (psi/in)	Positive TD (degrees F)	Design ESALs, millions							
				1.5	2	2.5	3	4	5	7.5	10
144	600	100	5	6.2	6.6	6.9	7.2	7.6	8.0	8.6	9.0
144	600	100	7	6.6	7.0	7.3	7.6	8.0	8.3	8.9	9.3
144	600	100	9	6.9	7.3	7.6	7.8	8.2	8.6	9.1	9.5
144	600	100	11	7.0	7.4	7.7	8.0	8.4	8.7	9.3	9.7
144	600	100	13	7.2	7.6	7.9	8.2	8.6	8.9	9.5	9.9
144	600	250	5	—	—	6.2	6.6	7.1	7.5	8.3	8.8
144	600	250	7	6.1	6.6	7.0	7.3	7.8	8.2	8.9	9.4
144	600	250	9	6.6	7.1	7.5	7.8	8.3	8.7	9.4	9.9
144	600	250	11	6.8	7.3	7.7	8.1	8.6	9.0	9.7	10.2
144	600	250	13	7.2	7.7	8.1	8.5	9.0	9.4	10.1	10.6
144	600	500	5	—	—	—	—	6.2	6.7	7.6	8.2
144	600	500	7	—	—	—	—	6.6	7.1	8.1	8.8
144	600	500	9	—	6.1	6.6	7.0	7.7	8.2	9.1	9.8
144	600	500	11	6.4	7.0	7.5	7.9	8.5	9.0	9.9	10.5
144	600	500	13	6.3	7.0	7.5	8.0	8.7	9.2	10.2	10.9
144	700	100	5	—	6.1	6.4	6.7	7.1	7.4	8.0	8.4
144	700	100	7	6.1	6.5	6.8	7.1	7.5	7.8	8.3	8.7
144	700	100	9	6.4	6.8	7.1	7.4	7.7	8.0	8.6	9.0
144	700	100	11	6.7	7.0	7.3	7.6	7.9	8.2	8.8	9.1
144	700	100	13	6.9	7.3	7.5	7.8	8.1	8.4	9.0	9.3
144	700	250	5	—	—	—	6.0	6.5	6.9	7.6	8.1
144	700	250	7	—	6.1	6.5	6.8	7.3	7.7	8.3	8.8
144	700	250	9	—	6.2	6.6	6.9	7.4	7.8	8.5	9.0
144	700	250	11	6.5	6.9	7.3	7.6	8.1	8.4	9.1	9.6
144	700	250	13	6.9	7.3	7.7	8.0	8.5	8.8	9.5	9.9
144	700	500	5	—	—	—	—	—	6.1	6.9	7.5
144	700	500	7	—	—	—	—	6.1	6.6	7.5	8.1
144	700	500	9	—	—	6.3	6.7	7.3	7.7	8.6	9.1
144	700	500	11	6.0	6.6	7.0	7.4	8.0	8.4	9.2	9.8
144	700	500	13	6.2	6.8	7.2	7.6	8.2	8.7	9.6	10.2
144	800	100	5	—	—	6.1	6.3	6.7	7.0	7.5	7.9
144	800	100	7	—	6.2	6.4	6.7	7.0	7.3	7.8	8.2
144	800	100	9	6.1	6.5	6.7	7.0	7.3	7.6	8.1	8.5
144	800	100	11	6.4	6.7	7.0	7.2	7.6	7.8	8.3	8.7
144	800	100	13	6.6	6.9	7.2	7.4	7.8	8.0	8.5	8.9
144	800	250	5	—	—	—	—	6.0	6.4	7.1	7.6
144	800	250	7	—	—	6.1	6.4	6.8	7.2	7.8	8.3
144	800	250	9	—	6.0	6.3	6.6	7.1	7.5	8.1	8.6
144	800	250	11	6.1	6.6	6.9	7.2	7.6	8.0	8.6	9.0
144	800	250	13	6.5	7.0	7.3	7.6	8.0	8.3	8.9	9.4
144	800	500	5	—	—	—	—	—	—	6.4	6.9
144	800	500	7	—	—	—	—	—	6.1	6.9	7.5
144	800	500	9	—	—	—	6.3	6.8	7.2	8.0	8.5
144	800	500	11	—	6.2	6.6	7.0	7.5	7.9	8.7	9.2
144	800	500	13	6.0	6.5	6.9	7.3	7.8	8.3	9.0	9.6
192	600	100	5	6.6	7.0	7.3	7.6	8.0	8.3	8.9	9.3
192	600	100	7	7.0	7.4	7.7	8.0	8.4	8.7	9.3	9.7
192	600	100	9	7.2	7.6	8.0	8.2	8.6	8.9	9.5	9.9
192	600	100	11	7.5	7.9	8.2	8.5	8.9	9.2	9.8	10.2
192	600	100	13	7.8	8.2	8.5	8.7	9.1	9.4	10.0	10.4
192	600	250	5	6.0	6.5	6.9	7.2	7.7	8.1	8.8	9.3
192	600	250	7	6.8	7.2	7.6	7.9	8.4	8.8	9.5	10.0
192	600	250	9	7.1	7.6	8.0	8.3	8.8	9.2	10.0	10.5
192	600	250	11	7.7	8.2	8.5	8.9	9.4	9.7	10.4	10.9
192	600	250	13	8.1	8.6	9.0	9.3	9.8	10.1	10.8	11.3
192	600	500	5	—	—	6.0	6.4	7.0	7.5	8.3	8.9
192	600	500	7	—	6.1	6.6	7.0	7.7	8.2	9.1	9.8
192	600	500	9	6.6	7.3	7.7	8.1	8.8	9.3	10.1	10.8
192	600	500	11	7.0	7.6	8.2	8.6	9.3	9.8	10.7	11.4
192	600	500	13	7.8	8.4	9.0	9.4	10.0	10.5	11.5	12.1
192	700	100	5	6.1	6.5	6.8	7.1	7.5	7.8	8.3	8.7
192	700	100	7	6.6	7.0	7.3	7.5	7.9	8.2	8.7	9.1
192	700	100	9	6.9	7.3	7.6	7.8	8.2	8.5	9.0	9.4
192	700	100	11	7.2	7.6	7.9	8.1	8.5	8.7	9.3	9.6
192	700	100	13	7.4	7.8	8.1	8.3	8.7	9.0	9.5	9.8
192	700	250	5	—	6.1	6.4	6.7	7.2	7.6	8.3	8.7
192	700	250	7	6.0	6.5	6.9	7.2	7.7	8.1	8.8	9.3
192	700	250	9	6.8	7.3	7.7	7.9	8.4	8.8	9.4	9.9
192	700	250	11	7.3	7.8	8.1	8.4	8.9	9.2	9.9	10.3
192	700	250	13	7.7	8.2	8.5	8.8	9.3	9.6	10.3	10.7

37

1 in = 25.4 mm, 1 psi = 6.89 kPa, 1 psi/in = .271 kPa/mm, °C = (°F-32)/1.8

Table 21. Slab thickness computed for high-strength base and 90 percent reliability (continued).

E_b = 1 million psi [6890 MPa], R = 90 percent, S_o = 0.39, P2 = 2.5, 12-ft-wide [3.7-m wide] lanes with AC shoulders.
Computed thicknesses less than 6.0 in [152 mm] or greater than 15.0 in [381 mm] are not shown.

Joint Spacing (in)	Flexural Strength (psi)	Subgrade k (psi/in)	Positive TD (degrees F)	Design ESALs, millions							
				1.5	2	2.5	3	4	5	7.5	10
192	700	500	5	—	—	—	6.0	6.5	7.0	7.8	8.4
192	700	500	7	—	6.1	6.5	6.9	7.5	7.9	8.7	9.3
192	700	500	9	—	6.4	6.9	7.3	7.9	8.4	9.3	9.9
192	700	500	11	6.9	7.5	8.0	8.3	8.9	9.4	10.2	10.8
192	700	500	13	7.6	8.2	8.6	9.0	9.6	10.0	10.8	11.4
192	800	100	5	—	6.2	6.5	6.7	7.1	7.4	7.9	8.2
192	800	100	7	6.3	6.6	6.9	7.1	7.5	7.8	8.3	8.6
192	800	100	9	6.6	7.0	7.2	7.5	7.8	8.1	8.6	8.9
192	800	100	11	6.9	7.3	7.5	7.7	8.1	8.3	8.8	9.2
192	800	100	13	7.1	7.5	7.7	8.0	8.3	8.6	9.0	9.4
192	800	250	5	—	—	6.1	6.4	6.8	7.2	7.8	8.2
192	800	250	7	—	6.3	6.6	6.9	7.4	7.7	8.4	8.8
192	800	250	9	6.6	7.0	7.3	7.6	8.0	8.3	9.0	9.4
192	800	250	11	7.0	7.5	7.8	8.0	8.5	8.8	9.4	9.8
192	800	250	13	7.3	7.7	8.0	8.3	8.7	9.1	9.7	10.1
192	800	500	5	—	—	5.3	5.6	6.2	6.6	7.3	7.9
192	800	500	7	—	—	6.2	6.6	7.1	7.5	8.3	8.8
192	800	500	9	—	6.4	6.8	7.2	7.7	8.1	8.9	9.5
192	800	500	11	6.7	7.3	7.7	8.0	8.5	8.9	9.7	10.2
192	800	500	13	7.0	7.5	8.0	8.3	8.9	9.3	10.1	10.7
240	600	100	5	6.9	7.3	7.7	7.9	8.3	8.7	9.2	9.6
240	600	100	7	7.3	7.7	8.0	8.3	8.7	9.0	9.6	10.0
240	600	100	9	7.7	8.1	8.4	8.7	9.1	9.4	10.0	10.3
240	600	100	11	8.1	8.4	8.7	9.0	9.4	9.7	10.2	10.6
240	600	100	13	8.3	8.7	9.0	9.2	9.6	9.9	10.5	10.8
240	600	250	5	6.6	7.1	7.4	7.8	8.3	8.6	9.4	9.8
240	600	250	7	7.2	7.7	8.1	8.4	8.9	9.3	10.0	10.5
240	600	250	9	7.9	8.4	8.8	9.1	9.6	10.0	10.7	11.2
240	600	250	11	8.5	8.9	9.3	9.6	10.1	10.5	11.2	11.7
240	600	250	13	8.7	9.2	9.6	9.9	10.4	10.8	11.5	12.0
240	600	500	5	—	—	6.1	6.5	7.2	7.7	8.6	9.3
240	600	500	7	6.6	7.2	7.7	8.1	8.7	9.2	10.1	10.7
240	600	500	9	7.1	7.8	8.3	8.8	9.5	10.0	11.0	11.7
240	600	500	11	8.1	8.8	9.3	9.8	10.4	11.0	11.9	12.6
240	600	500	13	8.6	9.3	9.8	10.3	11.0	11.6	12.6	13.3
240	700	100	5	6.5	6.9	7.2	7.5	7.9	8.2	8.7	9.1
240	700	100	7	7.0	7.4	7.7	7.9	8.3	8.6	9.1	9.5
240	700	100	9	7.4	7.8	8.1	8.3	8.7	9.0	9.5	9.8
240	700	100	11	7.7	8.1	8.4	8.6	9.0	9.2	9.7	10.1
240	700	100	13	8.0	8.3	8.6	8.8	9.2	9.5	10.0	10.3
240	700	250	5	6.3	6.7	7.1	7.4	7.8	8.2	8.8	9.3
240	700	250	7	7.0	7.4	7.8	8.1	8.5	8.9	9.5	10.0
240	700	250	9	7.7	8.1	8.4	8.7	9.2	9.5	10.2	10.6
240	700	250	11	8.0	8.5	8.8	9.1	9.6	9.9	10.6	11.0
240	700	250	13	8.5	9.0	9.3	9.6	10.0	10.4	11.0	11.5
240	700	500	5	—	—	6.2	6.5	7.1	7.6	8.4	9.0
240	700	500	7	—	6.5	7.0	7.4	8.0	8.5	9.4	10.0
240	700	500	9	7.3	7.9	8.4	8.7	9.3	9.7	10.5	11.1
240	700	500	11	7.9	8.5	8.9	9.3	9.9	10.4	11.2	11.8
240	700	500	13	8.7	9.3	9.7	10.1	10.7	11.1	11.9	12.5
240	800	100	5	6.2	6.6	6.9	7.1	7.5	7.7	8.3	8.6
240	800	100	7	6.7	7.1	7.4	7.6	7.9	8.2	8.7	9.0
240	800	100	9	7.1	7.5	7.7	8.0	8.3	8.6	9.0	9.4
240	800	100	11	7.5	7.8	8.0	8.3	8.6	8.8	9.3	9.7
240	800	100	13	7.7	8.0	8.3	8.5	8.8	9.1	9.5	9.9
240	800	250	5	—	6.1	6.5	6.8	7.2	7.6	8.2	8.7
240	800	250	7	6.7	7.2	7.5	7.8	8.2	8.5	9.1	9.5
240	800	250	9	7.2	7.6	7.9	8.2	8.6	9.0	9.6	10.0
240	800	250	11	7.8	8.2	8.5	8.8	9.2	9.5	10.1	10.5
240	800	250	13	8.2	8.7	9.0	9.2	9.6	10.0	10.5	10.9
240	800	500	5	—	—	6.0	6.3	6.9	7.3	8.1	8.6
240	800	500	7	6.1	6.7	7.1	7.4	8.0	8.4	9.1	9.7
240	800	500	9	7.2	7.7	8.1	8.5	9.0	9.4	10.1	10.6
240	800	500	11	7.8	8.4	8.8	9.1	9.6	10.0	10.8	11.3
240	800	500	13	8.5	9.0	9.4	9.8	10.3	10.7	11.4	11.9

1 in = 25.4 mm, 1 psi = 6.89 kPa, 1 psi/in = .271 kPa/mm, °C = (°F-32)/1.8

Table 22. Slab thickness computed for granular base and 85 percent reliability.

E_b = 25 ksi [172.25 MPa], R = 85 percent, S_o = 0.39, P2 = 2.5, 12-ft-wide [3.7-m wide] lanes with AC shoulders.
Computed thicknesses less than 6.0 in [152 mm] or greater than 15.0 in [381 mm] are not shown.

Joint Spacing (in)	Flexural Strength (psi)	Subgrade k (psi/in)	Positive TD (degrees F)	Design ESALs, millions							
				1.5	2	2.5	3	3.5	4	4.5	5
144	600	100	5	7.9	8.3	8.7	9.0	9.2	9.4	9.6	9.8
144	600	100	7	7.8	8.2	8.6	8.9	9.1	9.4	9.5	9.7
144	600	100	9	7.7	8.1	8.5	8.8	9.1	9.3	9.5	9.6
144	600	100	11	7.6	8.1	8.4	8.7	9.0	9.2	9.4	9.6
144	600	100	13	7.5	8.0	8.4	8.7	8.9	9.1	9.3	9.5
144	600	250	5	6.9	7.4	7.8	8.1	8.4	8.7	8.9	9.1
144	600	250	7	6.9	7.5	7.9	8.2	8.5	8.8	9.0	9.2
144	600	250	9	6.9	7.5	7.9	8.3	8.6	8.9	9.1	9.3
144	600	250	11	7.0	7.6	8.0	8.4	8.7	8.9	9.2	9.4
144	600	250	13	7.0	7.6	8.1	8.4	8.7	9.0	9.2	9.4
144	600	500	5	—	—	6.1	6.6	6.9	7.3	7.6	7.8
144	600	500	7	—	—	6.3	6.8	7.2	7.6	7.9	8.1
144	600	500	9	—	—	6.5	7.0	7.5	7.8	8.1	8.4
144	600	500	11	—	6.1	6.8	7.3	7.7	8.1	8.5	8.8
144	600	500	13	—	6.4	7.0	7.6	8.0	8.4	8.8	9.1
144	700	100	5	7.1	7.5	7.8	8.1	8.3	8.5	8.7	8.9
144	700	100	7	7.0	7.4	7.8	8.0	8.3	8.5	8.7	8.8
144	700	100	9	6.9	7.3	7.7	8.0	8.2	8.4	8.6	8.8
144	700	100	11	6.8	7.3	7.6	7.9	8.1	8.4	8.5	8.7
144	700	100	13	6.8	7.2	7.6	7.9	8.1	8.3	8.5	8.6
144	700	250	5	—	6.4	6.8	7.1	7.4	7.6	7.8	8.0
144	700	250	7	6.1	6.7	7.0	7.4	7.6	7.9	8.1	8.3
144	700	250	9	6.2	6.7	7.1	7.4	7.7	8.0	8.2	8.4
144	700	250	11	6.3	6.8	7.2	7.5	7.8	8.0	8.3	8.5
144	700	250	13	6.3	6.9	7.3	7.6	7.9	8.1	8.3	8.5
144	700	500	5	—	—	—	—	6.1	6.4	6.6	6.9
144	700	500	7	—	—	—	6.0	6.4	6.7	7.0	7.2
144	700	500	9	—	—	—	6.1	6.5	6.8	7.1	7.4
144	700	500	11	—	—	—	6.4	6.8	7.1	7.4	7.7
144	700	500	13	—	—	6.2	6.7	7.1	7.4	7.7	8.0
144	800	100	5	6.4	6.8	7.2	7.4	7.6	7.8	8.0	8.1
144	800	100	7	6.4	6.8	7.1	7.4	7.6	7.8	7.9	8.1
144	800	100	9	6.4	6.8	7.1	7.3	7.6	7.7	7.9	8.1
144	800	100	11	6.3	6.7	7.1	7.3	7.5	7.7	7.9	8.0
144	800	100	13	6.3	6.7	7.0	7.3	7.5	7.7	7.8	8.0
144	800	250	5	—	—	6.2	6.5	6.7	6.9	7.1	7.3
144	800	250	7	—	—	6.2	6.5	6.8	7.0	7.2	7.4
144	800	250	9	—	—	6.3	6.6	6.9	7.1	7.3	7.5
144	800	250	11	—	6.0	6.4	6.7	7.0	7.2	7.4	7.6
144	800	250	13	—	6.1	6.5	6.8	7.1	7.3	7.5	7.7
144	800	500	5	—	—	—	—	—	—	—	6.1
144	800	500	7	—	—	—	—	—	—	6.2	6.4
144	800	500	9	—	—	—	—	—	6.2	6.4	6.7
144	800	500	11	—	—	—	—	6.2	6.5	6.8	7.0
144	800	500	13	—	—	—	6.1	6.5	6.8	7.0	7.3
192	600	100	5	8.0	8.5	8.8	9.1	9.4	9.6	9.8	10.0
192	600	100	7	8.0	8.4	8.8	9.1	9.3	9.5	9.7	9.9
192	600	100	9	7.9	8.4	8.7	9.0	9.3	9.5	9.7	9.9
192	600	100	11	7.9	8.3	8.7	9.0	9.3	9.5	9.7	9.8
192	600	100	13	7.8	8.3	8.7	9.0	9.2	9.4	9.6	9.8
192	600	250	5	7.1	7.6	8.0	8.4	8.7	8.9	9.1	9.3
192	600	250	7	7.2	7.8	8.2	8.5	8.8	9.1	9.3	9.5
192	600	250	9	7.3	7.9	8.3	8.7	9.0	9.3	9.5	9.7
192	600	250	11	7.4	8.0	8.5	8.8	9.1	9.4	9.6	9.8
192	600	250	13	7.5	8.1	8.6	8.9	9.3	9.5	9.8	10.0
192	600	500	5	—	—	6.4	6.8	7.2	7.6	7.9	8.1
192	600	500	7	—	6.1	6.7	7.2	7.7	8.0	8.3	8.6
192	600	500	9	—	6.5	7.1	7.7	8.1	8.5	8.8	9.1
192	600	500	11	6.0	6.3	7.5	8.1	8.5	8.9	9.3	9.6
192	600	500	13	6.3	7.2	7.9	8.5	9.0	9.4	9.8	10.1
192	700	100	5	7.2	7.6	8.0	8.3	8.5	8.7	8.9	9.0
192	700	100	7	7.2	7.6	8.0	8.2	8.5	8.7	8.9	9.0
192	700	100	9	7.2	7.6	7.9	8.2	8.5	8.7	8.8	9.0
192	700	100	11	7.2	7.6	7.9	8.2	8.4	8.7	8.8	9.0
192	700	100	13	7.1	7.6	7.9	8.2	8.4	8.6	8.8	9.0
192	700	250	5	6.3	6.8	7.2	7.5	7.8	8.0	8.2	8.4
192	700	250	7	6.5	7.0	7.4	7.7	8.0	8.2	8.4	8.6
192	700	250	9	6.6	7.1	7.5	7.9	8.2	8.4	8.6	8.8
192	700	250	11	6.7	7.3	7.7	8.0	8.3	8.6	8.8	9.0
192	700	250	13	6.9	7.4	7.8	8.2	8.5	8.7	8.9	9.1

1 in = 25.4 mm, 1 psi = 6.89 kPa, 1 psi/in = .271 kPa/mm, °C = (°F-32)/1.8

Table 22. Slab thickness computed for granular base and 85 percent reliability (continued).

E_b = 25 ksi [172.25 MPa], R = 85 percent, S_o = 0.39, P2 = 2.5, 12-ft-wide [3.7-m wide] lanes with AC shoulders.

Computed thicknesses less than 6.0 in [152 mm] or greater than 15.0 in [381 mm] are not shown.

Joint Spacing (in)	Flexural Strength (psi)	Subgrade k (psi/in)	Positive TD (degrees F)	Design ESALs, millions							
				1.5	2	2.5	3	3.5	4	4.5	5
192	700	500	5	—	—	—	6.1	6.5	6.8	7.1	7.3
192	700	500	7	—	—	6.0	6.4	6.8	7.1	7.4	7.7
192	700	500	9	—	—	6.4	6.9	7.3	7.6	7.9	8.2
192	700	500	11	—	6.2	6.8	7.3	7.7	8.1	8.4	8.7
192	700	500	13	—	6.6	7.2	7.7	8.1	8.5	8.8	9.1
192	800	100	5	6.6	7.0	7.3	7.6	7.8	8.0	8.1	8.3
192	800	100	7	6.6	7.0	7.3	7.6	7.8	8.0	8.2	8.3
192	800	100	9	6.6	7.0	7.3	7.6	7.8	8.0	8.2	8.3
192	800	100	11	6.6	7.0	7.3	7.6	7.8	8.0	8.2	8.3
192	800	100	13	6.6	7.0	7.3	7.6	7.8	8.0	8.2	8.3
192	800	250	5	5.6	6.0	6.4	6.7	7.0	7.2	7.4	7.6
192	800	250	7	—	6.2	6.6	6.9	7.2	7.4	7.6	7.8
192	800	250	9	—	6.4	6.7	7.1	7.3	7.6	7.8	8.0
192	800	250	11	6.2	6.7	7.1	7.4	7.7	7.9	8.1	8.3
192	800	250	13	6.3	6.8	7.2	7.5	7.8	8.0	8.2	8.4
192	800	500	5	—	—	—	—	—	6.1	6.3	6.5
192	800	500	7	—	—	—	—	6.3	6.6	6.8	7.1
192	800	500	9	—	—	—	6.3	6.6	6.9	7.2	7.4
192	800	500	11	—	—	6.3	6.7	7.1	7.4	7.7	7.9
192	800	500	13	—	6.2	6.7	7.1	7.5	7.8	8.1	8.3
240	600	100	5	8.2	8.6	9.0	9.3	9.5	9.8	9.9	10.1
240	600	100	7	8.2	8.6	9.0	9.3	9.5	9.8	9.9	10.1
240	600	100	9	8.2	8.6	9.0	9.3	9.6	9.8	10.0	10.1
240	600	100	11	8.2	8.7	9.0	9.3	9.6	9.8	10.0	10.1
240	600	100	13	8.2	8.7	9.0	9.3	9.6	9.8	10.0	10.1
240	600	250	5	7.3	7.9	8.3	8.6	8.9	9.2	9.4	9.6
240	600	250	7	7.6	8.1	8.6	8.9	9.2	9.5	9.7	9.9
240	600	250	9	7.7	8.3	8.8	9.1	9.5	9.7	10.0	10.2
240	600	250	11	7.9	8.5	9.0	9.4	9.7	10.0	10.2	10.4
240	600	250	13	8.1	8.7	9.2	9.6	9.9	10.2	10.4	10.6
240	600	500	5	—	6.1	6.7	7.2	7.6	8.0	8.3	8.6
240	600	500	7	—	6.6	7.3	7.8	8.3	8.6	9.0	9.3
240	600	500	9	6.3	7.2	7.9	8.4	8.9	9.3	9.7	10.0
240	600	500	11	6.9	7.8	8.5	9.1	9.6	10.0	10.4	10.7
240	600	500	13	7.4	8.4	9.1	9.7	10.2	10.6	11.0	11.4
240	700	100	5	7.3	7.8	8.1	8.4	8.7	8.9	9.0	9.2
240	700	100	7	7.4	7.9	8.2	8.5	8.7	8.9	9.1	9.3
240	700	100	9	7.5	7.9	8.2	8.5	8.8	9.0	9.1	9.3
240	700	100	11	7.5	7.9	8.3	8.5	8.8	9.0	9.2	9.3
240	700	100	13	7.5	8.0	8.3	8.6	8.8	9.0	9.2	9.3
240	700	250	5	6.6	7.1	7.5	7.8	8.1	8.3	8.5	8.7
240	700	250	7	6.8	7.4	7.8	8.1	8.4	8.6	8.8	9.0
240	700	250	9	7.1	7.6	8.0	8.4	8.7	8.9	9.1	9.3
240	700	250	11	7.3	7.9	8.3	8.6	8.9	9.2	9.4	9.6
240	700	250	13	7.5	8.1	8.5	8.8	9.1	9.4	9.6	9.8
240	700	500	5	—	—	6.0	6.4	6.8	7.1	7.4	7.7
240	700	500	7	—	6.1	6.7	7.1	7.5	7.9	8.2	8.4
240	700	500	9	6.0	6.8	7.3	7.8	8.2	8.6	8.9	9.1
240	700	500	11	6.6	7.3	7.9	8.4	8.8	9.2	9.5	9.8
240	700	500	13	7.1	7.9	8.5	9.0	9.4	9.8	10.1	10.4
240	800	100	5	6.7	7.2	7.5	7.8	8.0	8.2	8.3	8.5
240	800	100	7	6.8	7.2	7.5	7.8	8.0	8.2	8.4	8.5
240	800	100	9	6.9	7.3	7.6	7.9	8.1	8.3	8.5	8.6
240	800	100	11	6.9	7.3	7.7	7.9	8.1	8.3	8.5	8.7
240	800	100	13	7.0	7.4	7.7	8.0	8.2	8.4	8.5	8.7
240	800	250	5	5.9	6.3	6.7	7.0	7.3	7.5	7.7	7.9
240	800	250	7	6.3	6.8	7.2	7.5	7.7	8.0	8.2	8.3
240	800	250	9	6.6	7.0	7.4	7.8	8.0	8.2	8.5	8.6
240	800	250	11	6.8	7.3	7.7	8.0	8.3	8.5	8.7	8.9
240	800	250	13	7.0	7.5	7.9	8.2	8.5	8.7	8.9	9.1
240	800	500	5	—	—	—	6.1	6.4	6.7	6.9	7.1
240	800	500	7	—	—	6.2	6.6	7.0	7.3	7.5	7.8
240	800	500	9	—	6.4	6.9	7.3	7.6	8.0	8.2	8.5
240	800	500	11	6.2	6.9	7.4	7.9	8.2	8.6	8.8	9.1
240	800	500	13	6.7	7.4	8.0	8.4	8.8	9.1	9.4	9.6

1 in = 25.4 mm, 1 psi = 6.89 kPa, 1 psi/in = .271 kPa/mm, °C = (°F-32)/1.8

Table 23. Slab thickness computed for treated base and 85 percent reliability.

E_b = 500 ksi [3445 MPa], R = 85 percent, S_o = 0.39, P2 = 2.5, 12-ft-wide [3.7-m wide] lanes with AC shoulders.
Computed thicknesses less than 6.0 in [152 mm] or greater than 15.0 in [381 mm] are not shown.

Joint Spacing (in)	Flexural Strength (psi)	Subgrade k (psi/in)	Positive TD (degrees F)	Design ESALs, millions							
				1.5	2	2.5	3	3.5	4	4.5	5
144	600	100	5	6.6	7.0	7.4	7.6	7.9	8.1	8.3	8.4
144	600	100	7	6.7	7.2	7.5	7.8	8.0	8.2	8.4	8.5
144	600	100	9	6.9	7.3	7.6	7.9	8.1	8.3	8.5	8.7
144	600	100	11	7.0	7.4	7.7	8.0	8.2	8.4	8.6	8.8
144	600	100	13	7.1	7.5	7.8	8.1	8.3	8.5	8.7	8.8
144	600	250	5	—	—	6.3	6.6	6.9	7.2	7.4	7.6
144	600	250	7	—	6.3	6.7	7.1	7.4	7.6	7.8	8.0
144	600	250	9	6.1	6.7	7.1	7.4	7.7	8.0	8.2	8.4
144	600	250	11	6.4	7.0	7.4	7.7	8.0	8.3	8.5	8.7
144	600	250	13	6.7	7.2	7.6	8.0	8.3	8.5	8.8	9.0
144	600	500	5	—	—	—	—	—	—	—	6.2
144	600	500	7	—	—	—	—	6.1	6.4	6.7	7.0
144	600	500	9	—	—	6.0	6.5	6.9	7.2	7.5	7.7
144	600	500	11	—	—	6.0	6.5	7.0	7.3	7.7	7.9
144	600	500	13	—	6.2	6.8	7.3	7.7	8.0	8.4	8.6
144	700	100	5	6.0	6.4	6.7	6.9	7.2	7.4	7.5	7.7
144	700	100	7	6.2	6.6	6.9	7.2	7.4	7.6	7.7	7.9
144	700	100	9	6.4	6.8	7.1	7.3	7.5	7.7	7.9	8.0
144	700	100	11	6.5	6.9	7.2	7.4	7.7	7.8	8.0	8.1
144	700	100	13	6.6	7.0	7.3	7.5	7.8	7.9	8.1	8.2
144	700	250	5	—	—	—	6.1	6.4	6.6	6.8	7.0
144	700	250	7	—	—	6.1	6.4	6.7	6.9	7.2	7.3
144	700	250	9	—	6.1	6.5	6.8	7.1	7.3	7.5	7.7
144	700	250	11	—	6.4	6.8	7.1	7.4	7.6	7.8	8.0
144	700	250	13	6.2	6.7	7.1	7.4	7.6	7.9	8.1	8.3
144	700	500	5	—	—	—	—	—	—	—	—
144	700	500	7	—	—	—	—	—	—	6.1	6.4
144	700	500	9	—	—	—	—	6.2	6.5	6.8	7.0
144	700	500	11	—	—	—	6.0	6.4	6.7	7.0	7.3
144	700	500	13	—	—	6.3	6.7	7.1	7.4	7.7	7.9
144	800	100	5	—	—	6.2	6.4	6.7	6.8	7.0	7.1
144	800	100	7	—	6.1	6.4	6.6	6.8	7.0	7.2	7.3
144	800	100	9	—	6.2	6.5	6.8	7.0	7.2	7.3	7.4
144	800	100	11	6.1	6.5	6.8	7.0	7.2	7.4	7.5	7.6
144	800	100	13	6.2	6.6	6.9	7.1	7.3	7.5	7.6	7.7
144	800	250	5	—	—	—	—	—	6.1	6.3	6.4
144	800	250	7	—	—	—	6.0	6.3	6.5	6.7	6.9
144	800	250	9	—	—	6.0	6.3	6.6	6.8	7.0	7.2
144	800	250	11	—	6.0	6.3	6.6	6.9	7.1	7.3	7.5
144	800	250	13	—	6.2	6.6	6.9	7.1	7.4	7.6	7.7
144	800	500	5	—	—	—	—	—	—	—	—
144	800	500	7	—	—	—	—	—	—	—	—
144	800	500	9	—	—	—	—	—	6.0	6.3	6.5
144	800	500	11	—	—	—	—	6.0	6.2	6.5	6.7
144	800	500	13	—	—	—	6.2	6.6	6.9	7.1	7.3
192	600	100	5	6.8	7.2	7.6	7.9	8.1	8.3	8.5	8.6
192	600	100	7	7.0	7.5	7.8	8.1	8.3	8.5	8.7	8.8
192	600	100	9	7.2	7.6	8.0	8.2	8.5	8.7	8.8	9.0
192	600	100	11	7.4	7.8	8.1	8.4	8.6	8.8	9.0	9.1
192	600	100	13	7.5	7.9	8.3	8.5	8.8	8.9	9.1	9.3
192	600	250	5	—	6.3	6.7	7.0	7.3	7.6	7.8	8.0
192	600	250	7	6.3	6.8	7.2	7.6	7.9	8.1	8.3	8.5
192	600	250	9	6.7	7.2	7.7	8.0	8.3	8.6	8.8	9.0
192	600	250	11	7.1	7.6	8.0	8.4	8.7	8.9	9.2	9.4
192	600	250	13	6.9	7.5	8.0	8.4	8.7	9.0	9.2	9.4
192	600	500	5	—	—	—	—	—	6.3	6.5	6.8
192	600	500	7	—	—	6.1	6.5	6.9	7.2	7.5	7.8
192	600	500	9	—	—	6.4	6.9	7.3	7.7	8.0	8.3
192	600	500	11	—	6.7	7.3	7.8	8.2	8.6	8.9	9.2
192	600	500	13	6.5	7.3	7.9	8.4	8.9	9.2	9.5	9.8
192	700	100	5	6.3	6.7	7.0	7.3	7.5	7.7	7.8	8.0
192	700	100	7	6.5	6.9	7.2	7.5	7.7	7.9	8.1	8.2
192	700	100	9	6.7	7.1	7.4	7.7	7.9	8.1	8.2	8.4
192	700	100	11	6.9	7.3	7.6	7.9	8.1	8.2	8.4	8.6
192	700	100	13	7.1	7.4	7.7	8.0	8.2	8.4	8.5	8.7
192	700	250	5	—	—	6.1	6.5	6.7	7.0	7.2	7.4
192	700	250	7	—	6.3	6.7	7.0	7.3	7.5	7.7	7.9
192	700	250	9	6.3	6.8	7.2	7.5	7.7	8.0	8.2	8.4
192	700	250	11	6.6	7.1	7.5	7.8	8.1	8.3	8.5	8.7
192	700	250	13	6.8	7.3	7.7	8.0	8.3	8.6	8.8	9.0

1 in = 25.4 mm, 1 psi = 6.89 kPa, 1 psi/in = .271 kPa/mm, °C = (°F-32)/1.8

Table 23. Slab thickness computed for treated base and 85 percent reliability (continued).

E_b = 500 ksi [3445 MPa], R = 85 percent, S_o = 0.39, P2 = 2.5, 12-ft-wide [3.7-m wide] lanes with AC shoulders.

Computed thicknesses less than 6.0 in [152 mm] or greater than 15.0 in [381 mm] are not shown.

Joint Spacing (in)	Flexural Strength (psi)	Subgrade k (psi/in)	Positive TD (degrees F)	Design ESALs, millions							
				1.5	2	2.5	3	3.5	4	4.5	5
192	700	500	5	—	—	—	—	—	—	6.0	6.3
192	700	500	7	—	—	—	6.1	6.5	6.8	7.0	7.3
192	700	500	9	—	—	6.2	6.6	7.0	7.3	7.6	7.8
192	700	500	11	—	6.5	7.0	7.4	7.8	8.1	8.4	8.6
192	700	500	13	6.4	7.1	7.6	8.0	8.4	8.7	9.0	9.2
192	800	100	5	—	6.2	6.5	6.7	6.9	7.1	7.3	7.4
192	800	100	7	6.1	6.5	6.8	7.0	7.2	7.4	7.6	7.7
192	800	100	9	6.4	6.7	7.0	7.2	7.4	7.6	7.8	7.9
192	800	100	11	6.5	6.9	7.2	7.4	7.6	7.8	7.9	8.1
192	800	100	13	6.7	7.1	7.3	7.6	7.8	7.9	8.1	8.2
192	800	250	5	—	—	—	6.1	6.4	6.6	6.8	6.9
192	800	250	7	—	—	6.3	6.6	6.8	7.0	7.2	7.4
192	800	250	9	—	6.4	6.7	7.0	7.3	7.5	7.7	7.8
192	800	250	11	6.3	6.7	7.1	7.4	7.6	7.8	8.0	8.2
192	800	250	13	6.5	7.0	7.4	7.6	7.9	8.1	8.3	8.5
192	800	500	5	—	—	—	—	—	—	—	—
192	800	500	7	—	—	—	—	6.1	6.3	6.6	6.8
192	800	500	9	—	—	—	6.3	6.6	6.9	7.2	7.4
192	800	500	11	—	6.2	6.7	7.0	7.4	7.6	7.9	8.1
192	800	500	13	—	6.4	6.9	7.3	7.6	7.9	8.2	8.4
240	600	100	5	7.1	7.5	7.8	8.1	8.4	8.6	8.7	8.9
240	600	100	7	7.4	7.8	8.1	8.4	8.6	8.8	9.0	9.2
240	600	100	9	7.6	8.0	8.4	8.6	8.9	9.1	9.2	9.4
240	600	100	11	7.8	8.2	8.5	8.8	9.0	9.2	9.4	9.6
240	600	100	13	8.0	8.4	8.7	9.0	9.2	9.4	9.6	9.7
240	600	250	5	6.2	6.7	7.2	7.5	7.8	8.0	8.3	8.5
240	600	250	7	6.8	7.4	7.8	8.1	8.4	8.7	8.9	9.1
240	600	250	9	7.3	7.9	8.3	8.7	8.9	9.2	9.4	9.6
240	600	250	11	7.5	8.1	8.6	8.9	9.2	9.5	9.7	9.9
240	600	250	13	8.0	8.6	9.0	9.4	9.7	9.9	10.2	10.4
240	600	500	5	—	—	—	6.2	6.6	6.9	7.2	7.5
240	600	500	7	—	—	6.4	6.9	7.3	7.7	8.0	8.3
240	600	500	9	6.3	7.0	7.6	8.1	8.5	8.9	9.2	9.5
240	600	500	11	7.2	7.9	8.5	9.0	9.4	9.8	10.1	10.4
240	600	500	13	7.5	8.3	8.9	9.5	9.9	10.3	10.7	11.0
240	700	100	5	6.6	7.0	7.3	7.5	7.8	7.9	8.1	8.3
240	700	100	7	6.9	7.3	7.6	7.8	8.1	8.2	8.4	8.6
240	700	100	9	7.1	7.5	7.8	8.1	8.3	8.5	8.6	8.8
240	700	100	11	7.4	7.7	8.0	8.3	8.5	8.7	8.8	9.0
240	700	100	13	7.5	7.9	8.2	8.5	8.7	8.9	9.0	9.2
240	700	250	5	—	6.3	6.7	7.0	7.3	7.5	7.7	7.9
240	700	250	7	6.5	6.9	7.3	7.6	7.9	8.1	8.3	8.5
240	700	250	9	6.8	7.3	7.7	8.0	8.3	8.5	8.7	8.9
240	700	250	11	7.3	7.8	8.2	8.5	8.8	9.0	9.2	9.4
240	700	250	13	7.7	8.2	8.6	8.9	9.2	9.4	9.6	9.8
240	700	500	5	—	—	—	6.0	6.3	6.6	6.9	7.1
240	700	500	7	—	—	6.4	6.8	7.2	7.5	7.7	8.0
240	700	500	9	6.3	6.9	7.5	7.9	8.2	8.5	8.8	9.0
240	700	500	11	6.7	7.4	7.9	8.4	8.8	9.1	9.4	9.6
240	700	500	13	7.5	8.2	8.7	9.2	9.6	9.9	10.2	10.4
240	800	100	5	6.2	6.6	6.8	7.1	7.3	7.5	7.6	7.7
240	800	100	7	6.5	6.9	7.2	7.4	7.6	7.8	7.9	8.1
240	800	100	9	6.8	7.1	7.4	7.7	7.9	8.0	8.2	8.3
240	800	100	11	7.0	7.4	7.6	7.9	8.1	8.2	8.4	8.5
240	800	100	13	7.2	7.5	7.8	8.0	8.2	8.4	8.6	8.7
240	800	250	5	—	—	6.3	6.6	6.8	7.0	7.2	7.4
240	800	250	7	6.1	6.6	6.9	7.2	7.5	7.7	7.9	8.0
240	800	250	9	6.6	7.0	7.4	7.7	7.9	8.1	8.3	8.5
240	800	250	11	7.1	7.5	7.9	8.2	8.4	8.6	8.8	9.0
240	800	250	13	7.4	7.9	8.3	8.5	8.8	9.0	9.2	9.3
240	800	500	5	—	—	—	—	6.0	6.3	6.5	6.7
240	800	500	7	—	—	6.2	6.6	6.9	7.2	7.4	7.7
240	800	500	9	6.2	6.7	7.2	7.6	7.9	8.1	8.4	8.6
240	800	500	11	6.7	7.3	7.8	8.2	8.5	8.8	9.0	9.2
240	800	500	13	7.4	8.0	8.5	8.8	9.2	9.4	9.7	9.9

1 in = 25.4 mm, 1 psi = 6.89 kPa, 1 psi/in = .271 kPa/mm, °C = (°F-32)/1.8

Table 24. Slab thickness computed for high-strength base and 85 percent reliability.

E_b = 1 million psi [6890 MPa], R = 85 percent, S_o = 0.39, P2 = 2.5, 12-ft-wide [3.7-m wide] lanes with AC shoulders.
Computed thicknesses less than 6.0 in [152 mm] or greater than 15.0 in [381 mm] are not shown.

Joint Spacing (in)	Flexural Strength (psi)	Subgrade k (psi/in)	Positive TD (degrees F)	Design ESALs, millions							
				1.5	2	2.5	3	3.5	4	4.5	5
144	600	100	5	—	6.3	6.6	6.9	7.1	7.3	7.5	7.7
144	600	100	7	6.2	6.7	7.0	7.2	7.5	7.7	7.8	8.0
144	600	100	9	6.5	6.9	7.3	7.5	7.7	7.9	8.1	8.2
144	600	100	11	6.7	7.1	7.4	7.7	7.9	8.1	8.2	8.4
144	600	100	13	6.9	7.3	7.6	7.9	8.1	8.3	8.5	8.6
144	600	250	5	—	—	—	6.2	6.5	6.7	6.9	7.1
144	600	250	7	—	6.2	6.6	6.9	7.2	7.4	7.7	7.8
144	600	250	9	6.3	6.8	7.1	7.5	7.7	8.0	8.2	8.3
144	600	250	11	6.4	6.9	7.3	7.7	7.9	8.2	8.4	8.6
144	600	250	13	6.8	7.4	7.7	8.1	8.3	8.6	8.8	9.0
144	600	500	5	—	—	—	—	—	—	6.0	6.2
144	600	500	7	—	—	—	—	—	6.0	6.3	6.6
144	600	500	9	—	—	6.1	6.5	6.9	7.2	7.5	7.7
144	600	500	11	—	6.5	7.0	7.4	7.8	8.1	8.3	8.5
144	600	500	13	—	6.5	7.0	7.5	7.8	8.2	8.4	8.7
144	700	100	5	—	—	6.1	6.4	6.6	6.8	7.0	7.1
144	700	100	7	—	6.2	6.5	6.8	7.0	7.2	7.3	7.5
144	700	100	9	6.2	6.5	6.8	7.1	7.3	7.5	7.6	7.7
144	700	100	11	6.4	6.7	7.0	7.3	7.5	7.7	7.8	7.9
144	700	100	13	6.6	7.0	7.3	7.5	7.7	7.9	8.0	8.1
144	700	250	5	—	—	—	—	—	6.1	6.3	6.5
144	700	250	7	—	—	6.2	6.5	6.7	6.9	7.1	7.3
144	700	250	9	—	—	6.2	6.5	6.8	7.0	7.2	7.4
144	700	250	11	6.1	6.6	6.9	7.2	7.5	7.7	7.9	8.1
144	700	250	13	6.5	7.0	7.3	7.6	7.9	8.1	8.3	8.5
144	700	500	5	—	—	—	—	—	—	—	—
144	700	500	7	—	—	—	—	—	—	—	6.1
144	700	500	9	—	—	—	6.3	6.6	6.8	7.1	7.3
144	700	500	11	—	6.2	6.6	7.0	7.3	7.5	7.8	8.0
144	700	500	13	—	6.3	6.8	7.2	7.5	7.8	8.0	8.2
144	800	100	5	—	—	—	6.0	6.2	6.4	6.6	6.7
144	800	100	7	—	—	6.2	6.4	6.6	6.8	6.9	7.0
144	800	100	9	—	6.2	6.5	6.7	6.9	7.1	7.2	7.3
144	800	100	11	6.1	6.4	6.7	6.9	7.1	7.3	7.4	7.6
144	800	100	13	6.3	6.7	6.9	7.2	7.3	7.5	7.6	7.8
144	800	250	5	—	—	—	—	—	—	—	6.0
144	800	250	7	—	—	—	6.0	6.3	6.5	6.7	6.8
144	800	250	9	—	—	6.0	6.3	6.5	6.8	6.9	7.1
144	800	250	11	—	6.2	6.6	6.9	7.1	7.3	7.5	7.6
144	800	250	13	6.2	6.6	7.0	7.2	7.5	7.7	7.8	8.0
144	800	500	5	—	—	—	—	—	—	—	—
144	800	500	7	—	—	—	—	—	—	—	—
144	800	500	9	—	—	—	—	6.1	6.4	6.6	6.8
144	800	500	11	—	—	6.2	6.6	6.8	7.1	7.3	7.5
144	800	500	13	—	6.1	6.5	6.9	7.2	7.4	7.6	7.8
192	600	100	5	6.2	6.6	7.0	7.2	7.5	7.7	7.8	8.0
192	600	100	7	6.7	7.1	7.4	7.7	7.9	8.1	8.2	8.4
192	600	100	9	6.9	7.3	7.6	7.9	8.1	8.3	8.5	8.6
192	600	100	11	7.2	7.6	7.9	8.2	8.4	8.6	8.8	8.9
192	600	100	13	7.5	7.9	8.2	8.4	8.6	8.8	9.0	9.1
192	600	250	5	—	6.1	6.5	6.8	7.1	7.3	7.5	7.7
192	600	250	7	6.4	6.9	7.3	7.6	7.8	8.1	8.3	8.4
192	600	250	9	6.7	7.2	7.6	7.9	8.2	8.5	8.7	8.8
192	600	250	11	7.3	7.8	8.2	8.5	8.7	9.0	9.2	9.4
192	600	250	13	7.7	8.2	8.6	8.9	9.2	9.4	9.6	9.8
192	600	500	5	—	—	—	—	6.2	6.5	6.8	7.0
192	600	500	7	—	—	6.1	6.5	6.9	7.2	7.4	7.7
192	600	500	9	6.1	6.3	7.3	7.7	8.0	8.3	8.5	8.8
192	600	500	11	6.4	7.1	7.6	8.1	8.4	8.7	9.0	9.3
192	600	500	13	7.3	7.9	8.5	8.9	9.2	9.5	9.8	10.0
192	700	100	5	—	6.2	6.5	6.8	7.0	7.2	7.3	7.5
192	700	100	7	6.3	6.7	7.0	7.2	7.4	7.6	7.8	7.9
192	700	100	9	6.6	7.0	7.3	7.5	7.7	7.9	8.0	8.2
192	700	100	11	6.9	7.3	7.6	7.8	8.0	8.2	8.3	8.5
192	700	100	13	7.2	7.5	7.8	8.0	8.2	8.4	8.5	8.7
192	700	250	5	—	—	6.1	6.4	6.6	6.9	7.0	7.2
192	700	250	7	—	6.1	6.5	6.8	7.1	7.3	7.5	7.7
192	700	250	9	6.5	6.9	7.3	7.6	7.8	8.1	8.2	8.4
192	700	250	11	7.0	7.4	7.8	8.1	8.3	8.5	8.7	8.9
192	700	250	13	7.4	7.8	8.2	8.5	8.7	8.9	9.1	9.3

43

1 in = 25.4 mm, 1 psi = 6.89 kPa, 1 psi/in = .271 kPa/mm, °C = (°F-32)/1.8

Table 24. Slab thickness computed for high-strength base and 85 percent reliability (continued).

E_b = 1 million psi [6890 MPa], R = 85 percent, S_o = 0.39, P2 = 2.5, 12-ft-wide [3.7-m wide] lanes with AC shoulders.
Computed thicknesses less than 6.0 in [152 mm] or greater than 15.0 in [381 mm] are not shown.

Joint Spacing (in)	Flexural Strength (psi)	Subgrade k (psi/in)	Positive TD (degrees F)	Design ESALs, millions							
				1.5	2	2.5	3	3.5	4	4.5	5
192	700	500	5	—	—	—	—	—	6.1	6.3	6.6
192	700	500	7	—	—	6.1	6.4	6.7	7.0	7.2	7.5
192	700	500	9	—	—	6.4	6.8	7.1	7.4	7.7	7.9
192	700	500	11	6.5	7.1	7.5	7.9	8.2	8.5	8.7	8.9
192	700	500	13	7.2	7.7	8.2	8.6	8.9	9.1	9.4	9.6
192	800	100	5	—	—	6.2	6.4	6.6	6.8	6.9	7.1
192	800	100	7	6.0	6.3	6.6	6.8	7.0	7.2	7.4	7.5
192	800	100	9	6.4	6.7	7.0	7.2	7.4	7.5	7.7	7.8
192	800	100	11	6.7	7.0	7.3	7.5	7.7	7.8	8.0	8.1
192	800	100	13	6.9	7.2	7.5	7.7	7.9	8.0	8.2	8.3
192	800	250	5	—	—	—	6.0	6.3	6.5	6.7	6.8
192	800	250	7	—	—	6.3	6.6	6.8	7.0	7.2	7.4
192	800	250	9	6.2	6.7	7.0	7.3	7.5	7.7	7.9	8.0
192	800	250	11	6.7	7.1	7.5	7.7	8.0	8.1	8.3	8.5
192	800	250	13	6.9	7.4	7.7	8.0	8.2	8.4	8.6	8.7
192	800	500	5	—	—	—	—	—	—	—	6.2
192	800	500	7	—	—	—	6.1	6.4	6.7	6.9	7.1
192	800	500	9	—	6.0	6.4	6.8	7.1	7.3	7.5	7.7
192	800	500	11	6.3	6.9	7.3	7.6	7.9	8.1	8.4	8.5
192	800	500	13	6.6	7.1	7.6	7.9	8.2	8.5	8.7	8.9
240	600	100	5	6.6	7.0	7.3	7.6	7.8	8.0	8.2	8.3
240	600	100	7	7.0	7.4	7.7	8.0	8.2	8.4	8.6	8.7
240	600	100	9	7.4	7.8	8.1	8.4	8.6	8.8	8.9	9.1
240	600	100	11	7.8	8.1	8.5	8.7	8.9	9.1	9.2	9.4
240	600	100	13	8.0	8.4	8.7	8.9	9.2	9.3	9.5	9.6
240	600	250	5	6.2	6.7	7.1	7.4	7.7	7.9	8.1	8.3
240	600	250	7	6.8	7.3	7.7	8.0	8.3	8.5	8.7	8.9
240	600	250	9	7.5	8.0	8.4	8.7	9.0	9.2	9.4	9.6
240	600	250	11	8.1	8.6	8.9	9.3	9.5	9.7	9.9	10.1
240	600	250	13	8.3	8.8	9.2	9.5	9.8	10.0	10.2	10.4
240	600	500	5	—	—	—	6.0	6.3	6.6	6.9	7.2
240	600	500	7	6.1	6.7	7.2	7.6	7.9	8.2	8.5	8.7
240	600	500	9	6.6	7.3	7.8	8.3	8.6	8.9	9.2	9.5
240	600	500	11	7.6	8.3	8.8	9.3	9.6	9.9	10.2	10.5
240	600	500	13	8.0	8.7	9.3	9.7	10.1	10.5	10.8	11.0
240	700	100	5	6.3	6.6	6.9	7.2	7.4	7.6	7.7	7.9
240	700	100	7	6.7	7.1	7.4	7.6	7.8	8.0	8.2	8.3
240	700	100	9	7.1	7.5	7.8	8.0	8.2	8.4	8.5	8.7
240	700	100	11	7.5	7.8	8.1	8.3	8.5	8.7	8.8	9.0
240	700	100	13	7.7	8.1	8.3	8.6	8.8	8.9	9.1	9.2
240	700	250	5	—	6.4	6.7	7.0	7.3	7.5	7.7	7.8
240	700	250	7	6.6	7.1	7.4	7.7	8.0	8.2	8.4	8.5
240	700	250	9	7.3	7.8	8.1	8.4	8.6	8.8	9.0	9.2
240	700	250	11	7.6	8.1	8.5	8.7	9.0	9.2	9.4	9.6
240	700	250	13	8.2	8.6	9.0	9.2	9.5	9.7	9.9	10.0
240	700	500	5	—	—	—	6.1	6.4	6.7	6.9	7.1
240	700	500	7	—	6.0	6.5	6.9	7.2	7.5	7.8	8.0
240	700	500	9	6.9	7.5	7.9	8.3	8.6	8.9	9.1	9.3
240	700	500	11	7.4	8.0	8.5	8.8	9.2	9.4	9.7	9.9
240	700	500	13	8.3	8.8	9.3	9.7	10.0	10.2	10.5	10.7
240	800	100	5	6.0	6.3	6.6	6.8	7.0	7.2	7.3	7.5
240	800	100	7	6.5	6.8	7.1	7.3	7.5	7.7	7.8	7.9
240	800	100	9	6.9	7.2	7.5	7.7	7.9	8.0	8.2	8.3
240	800	100	11	7.2	7.5	7.8	8.0	8.2	8.3	8.5	8.6
240	800	100	13	7.4	7.8	8.0	8.2	8.4	8.6	8.7	8.8
240	800	250	5	—	—	6.1	6.4	6.7	6.9	7.1	7.2
240	800	250	7	6.4	6.8	7.2	7.4	7.7	7.9	8.0	8.2
240	800	250	9	6.8	7.3	7.6	7.9	8.1	8.3	8.5	8.6
240	800	250	11	7.5	7.9	8.2	8.5	8.7	8.9	9.1	9.2
240	800	250	13	7.9	8.3	8.7	8.9	9.1	9.3	9.5	9.6
240	800	500	5	—	—	—	—	6.2	6.5	6.7	6.9
240	800	500	7	—	6.3	6.7	7.0	7.3	7.6	7.8	8.0
240	800	500	9	6.8	7.3	7.7	8.1	8.3	8.6	8.8	9.0
240	800	500	11	7.4	8.0	8.4	8.7	9.0	9.2	9.4	9.6
240	800	500	13	8.1	8.7	9.0	9.4	9.6	9.9	10.1	10.3

1 in = 25.4 mm, 1 psi = 6.89 kPa, 1 psi/in = .271 kPa/mm, °C = (°F-32)/1.8

Example Determination of Required Slab Thickness. Using the rigid pavement design equations presented previously, a slab thickness that is adequate to support the design ESALs must be determined by iteration. Consider, for example, the following inputs:

Input	Value
Estimated future traffic, W_{18}	20 million
Design reliability, R	95 percent
Overall standard deviation, S_o	0.39
Design serviceability loss, $\Delta PSI = P_1 - P_2$	4.5 - 2.5 = 2.0
Effective subgrade k-value, k	100 psi/in [27 kPa/mm]
Mean concrete modulus of rupture, S'_c	700 psi [4827 kPa]
Mean concrete elastic modulus, E_c	$26454\ S'^{0.77}_c$ = 4,100,000 psi [28,270 MPa]
Joint spacing, L	16 ft = 192 in [4.88 m]
Base modulus, E_b	1,000,000 psi [6895 MPa] (high-strength base)
Slab/base friction coefficient, f	35
Base thickness, H_b	5 in [127 mm]
Average annual wind speed	10 mph [16 km/h]
Average annual temperature	53°F [11.7°C]
Average annual precipitation	40 in [1016 mm]
Lane edge support condition	Conventional slab width (12 ft [3.66 m]) and AC shoulders

For the above climatic inputs and a trial slab thickness of 11 in [279 mm], an effective positive temperature differential TD of 9°F [5°C] was computed from Equation 48. Using the rigid pavement design equations, a slab thickness of 10.75 in [273 mm] was found to be needed for a design traffic level of 20 million ESALs and a design reliability level of 95 percent. Similarly, Table 18 indicates a required slab thickness of 10.7 in [272 mm]. These thicknesses are close to the initial estimate of 11 in [279 mm]. If the thickness obtained differs by an inch or more from the estimated thickness used to compute the effective positive temperature differential, the determination of the required slab thickness should be repeated, beginning with a new effective temperature differential for the new trial slab thickness.

Design Check for Joint Load Position Cracking. This check is not necessary if dowels are to be used at the transverse joints. Dowels reduce the stresses at the joint to levels much lower than those at the midslab load position. Cracking near adequately doweled joints is uncommon, and when it does occur, is attributable to causes other than fatigue damage.

If dowels are not used at the transverse joints, a check must be made to ensure that stresses created at the top of the slab when the axle load is at the joint are not excessive. Under certain design and climatic conditions, truck axle loadings near an undoweled transverse joint may produce higher tensile stresses at the top of the slab than the stresses produced at the bottom of the slab by midslab loading. These repeated high tensile stresses could result in the development of corner breaks or diagonal cracks. The load and climatic conditions that could potentially contribute to the critical stress being produced by joint loading are described below.

Axle load stress. When the axle load is near the transverse joint, a tensile stress occurs at the top of the slab.

Negative temperature differential stress. Negative (nighttime) temperature differentials cause corners to curl upward, which, due to the weight of the slab, produces a tensile stress at the slab surface.

Construction curling stress. If a high positive temperature differential through the slab exists in a concrete slab when it hardens (at which time the slab is flat), upward corner and edge curling may occur shortly thereafter when the temperature gradient dissipates. A high positive temperature differential occurs particularly on sunny days and when conventional curing procedures are used. This temperature differential has not been measured extensively and its typical magnitude is not well known at the present time.

Moisture gradient stress. Moisture shrinkage warping of the top of the slab occurs over time. The stress induced by this type of warping can be determined by representing the moisture warping by an equivalent temperature gradient.

It is difficult to quantify construction curling stress and moisture gradient stress separately. However, their combined effect can be thought of as the positive temperature differential required to bring the slab into a flat position in the absence of an actual temperature differential through the slab. An approximate equivalent temperature differential may be assumed that is related to the climate of the site and to conventional curing procedures (i.e., curing compound, no wet cure):

> **Wet climate** (Annual precipitation $\geq$ 30 in [762 mm] or Thornthwaite Moisture Index > 0): 0 to 2°F per inch [0 to 0.044°C per mm] of slab thickness.

> **Dry climate** (Annual precipitation < 30 in [762 mm] or Thornthwaite Moisture Index < 0): 1 to 3°F per inch [0.022 to 0.066°C per mm] of slab thickness.

If wet curing or night construction are used, these values may be reduced significantly.

The procedure to check for critical stress for the joint loading position for pavements without mechanical load transfer devices equivalent to dowel bars consists of the following steps:

1. Determine the required slab thickness as described previously, assuming that the midslab loading position is critical.
2. Compute the midslab stress for the required slab thickness and the site's climatic conditions.
3. Estimate a total equivalent negative temperature differential that considers the contributions of the effective (weighted average annual) negative temperature differential, construction temperature differential, and moisture differential.
4. Estimate the critical stress at the top of the slab due to joint loading and the total equivalent negative temperature differential.
5. Compare the midslab loading stress (Step 2) with the joint loading stress (Step 4). If the joint loading position yields a stress equal to or higher than the midslab loading position, consideration should be given to redesign of the joints to reduce the joint loading stress.

Step 1. Determine the Required Slab Thickness assuming that the midslab loading position is critical, using the design equations or tables provided earlier.   Note that the effect of slab/base friction is included in the required slab thickness obtained by either of these methods.

Step 2. Compute the Midslab Stress for the required slab thickness and the site's effective positive temperature differential. This may already have been done in Step 1 if the required slab thickness was determined using the design equations provided earlier. The midslab stress may also be estimated by the following method:

(a) Use the charts provided in Figures 48 through 53 to determine, interpolating as necessary, the midslab stress assuming full friction between the slab and base. Charts are provided for two levels of base modulus and three levels of subgrade k.
(b) Use Equation 46 to compute a friction adjustment factor.
(c) Multiply the full friction stress by the friction adjustment factor to obtain the proper estimate of the midslab stress.

Step 3. Estimate the Total Equivalent Negative Temperature Differential from the following sources.

(a) Effective negative temperature differential from the following equation:

$$\textit{effective negative TD} \;=\; -18.14 \;+\; \frac{52.01}{D} \;+\; 0.394 \; \textit{WIND}$$

$$+ \; 0.07 \; \textit{TEMP} \;+\; 0.00407 \; \textit{PRECIP}$$

[51]

where effective negative TD = top temperature minus bottom temperature, °F
D = slab thickness, inches
WIND = mean annual wind speed, mph
TEMP = mean annual temperature, °F
PRECIP = mean annual precipitation, inches

(b) Combined moisture gradient and construction temperature differential:

Wet climate (Annual precipitation $\geq$ 30 in [762 mm] or Thornthwaite Moisture Index $>$ 0): 0 to 2°F per inch [0 to 0.044°C per mm] of slab thickness.

Dry climate (Annual precipitation $<$ 30 in [762 mm] or Thornthwaite Moisture Index $<$ 0): 1 to 3°F per inch [0.022 to 0.066°C per mm] of slab thickness.

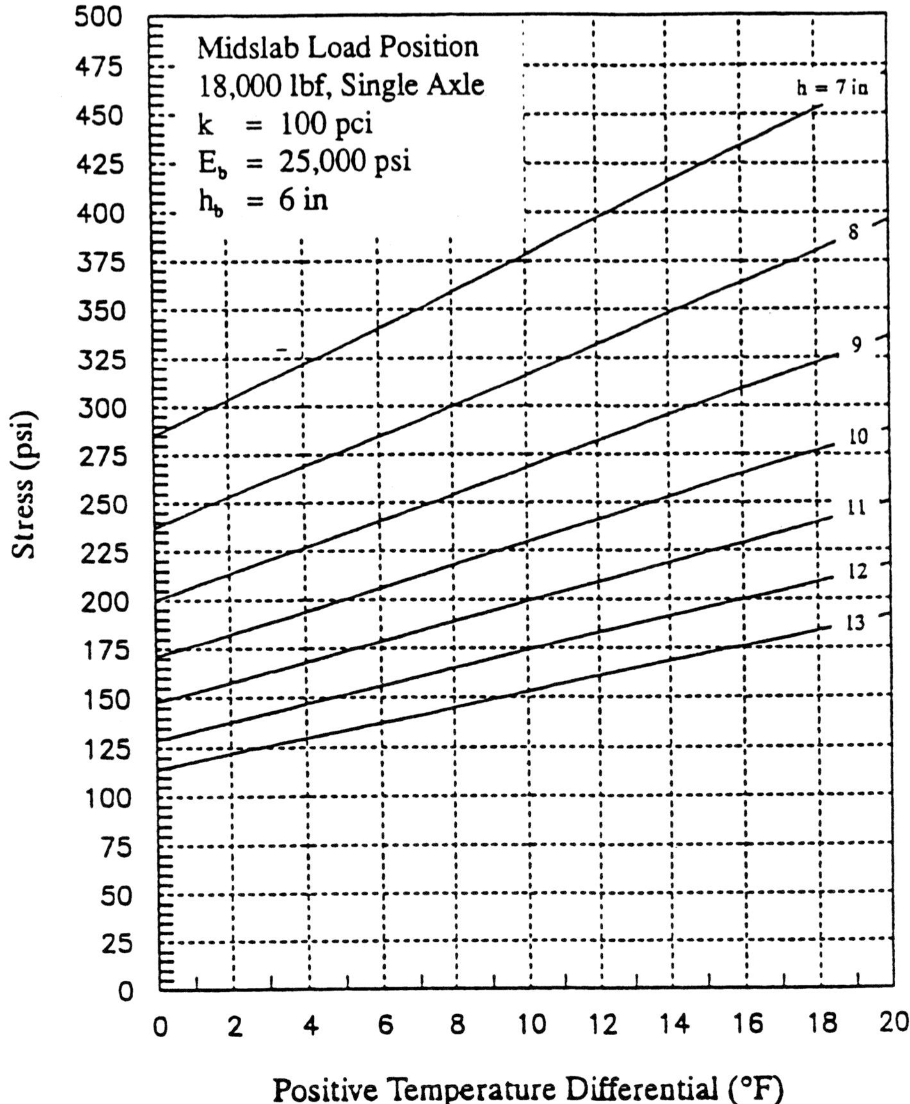

1 lbf = 4.45 N, 1 pci = 0.271 kPa/mm, 1 psi = 6.89 kPa, 1 in = 25.4 mm, °C = (°F - 32)/1.8

Figure 48. Tensile stress at bottom of slab for midslab loading position, positive temperature differential, and full friction, for aggregate base and soft subgrade.

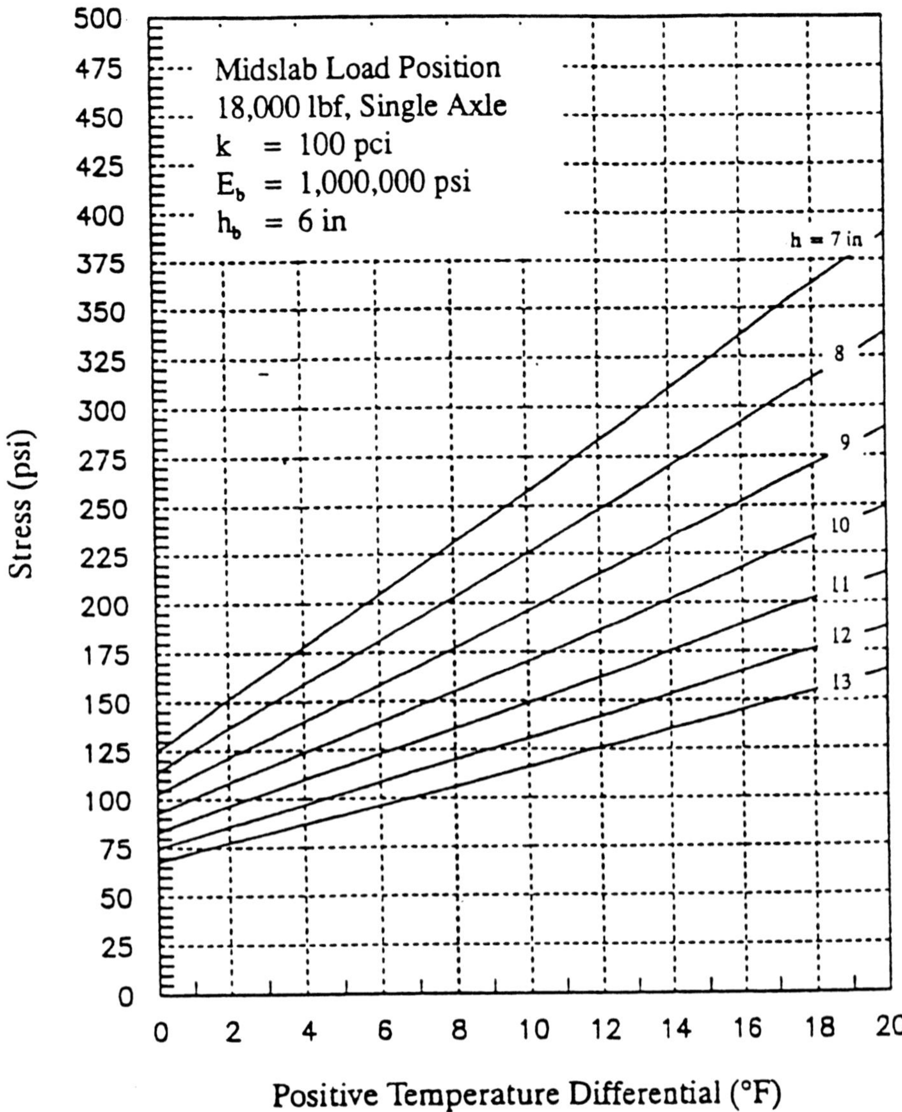

1 lbf = 4.45 N, 1 pci = 0.271 kPa/mm, 1 psi = 6.89 kPa, 1 in = 25.4 mm, °C = (°F - 32)/1.8

Figure 49. Tensile stress at bottom of slab for midslab loading position, positive temperature differential, and full friction, for high-strength base and soft subgrade.

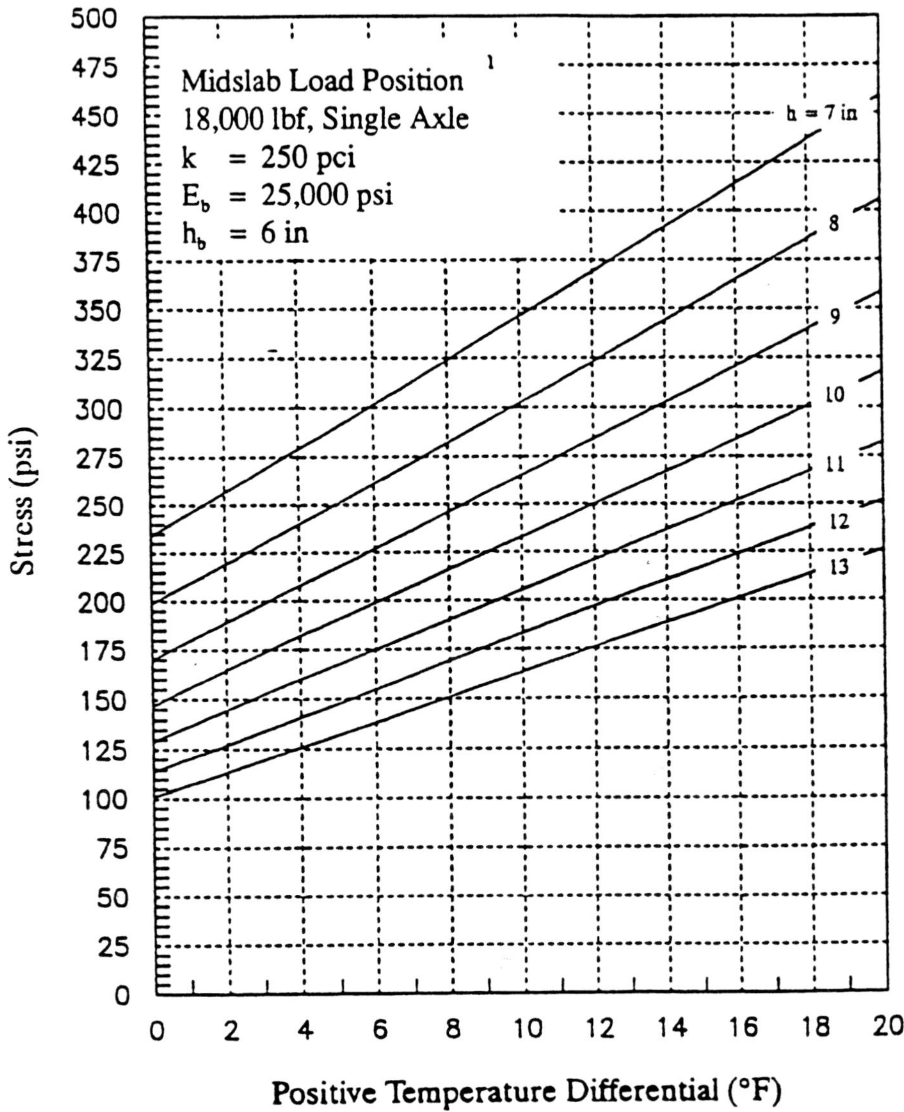

1 lbf = 4.45 N, 1 pci = 0.271 kPa/mm, 1 psi = 6.89 kPa, 1 in = 25.4 mm, °C = (°F - 32)/1.8

Figure 50. Tensile stress at bottom of slab for midslab loading position, positive temperature differential, and full friction, for aggregate base and medium subgrade.

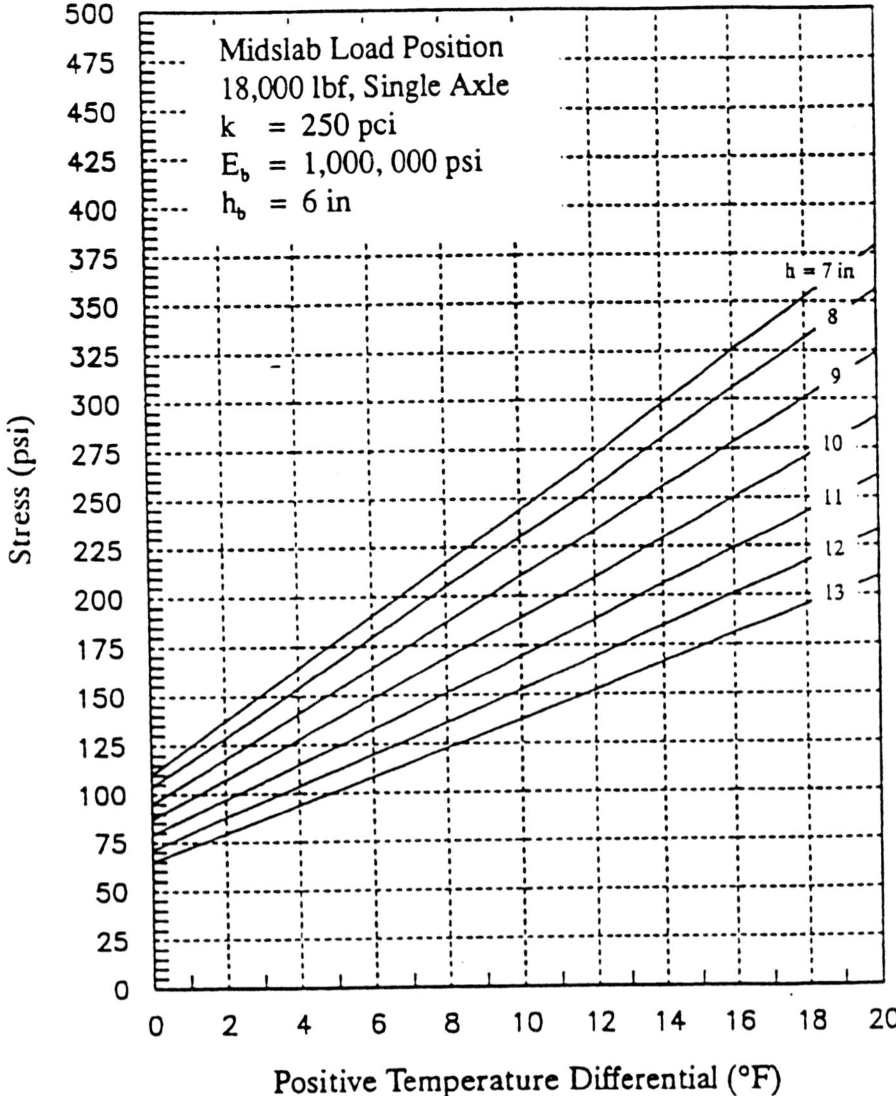

1 lbf = 4.45 N, 1 pci = 0.271 kPa/mm, 1 psi = 6.89 kPa, 1 in = 25.4 mm, °C = (°F - 32)/1.8

Figure 51. Tensile stress at bottom of slab for midslab loading position, positive temperature differential, and full friction, for high-strength base and medium subgrade.

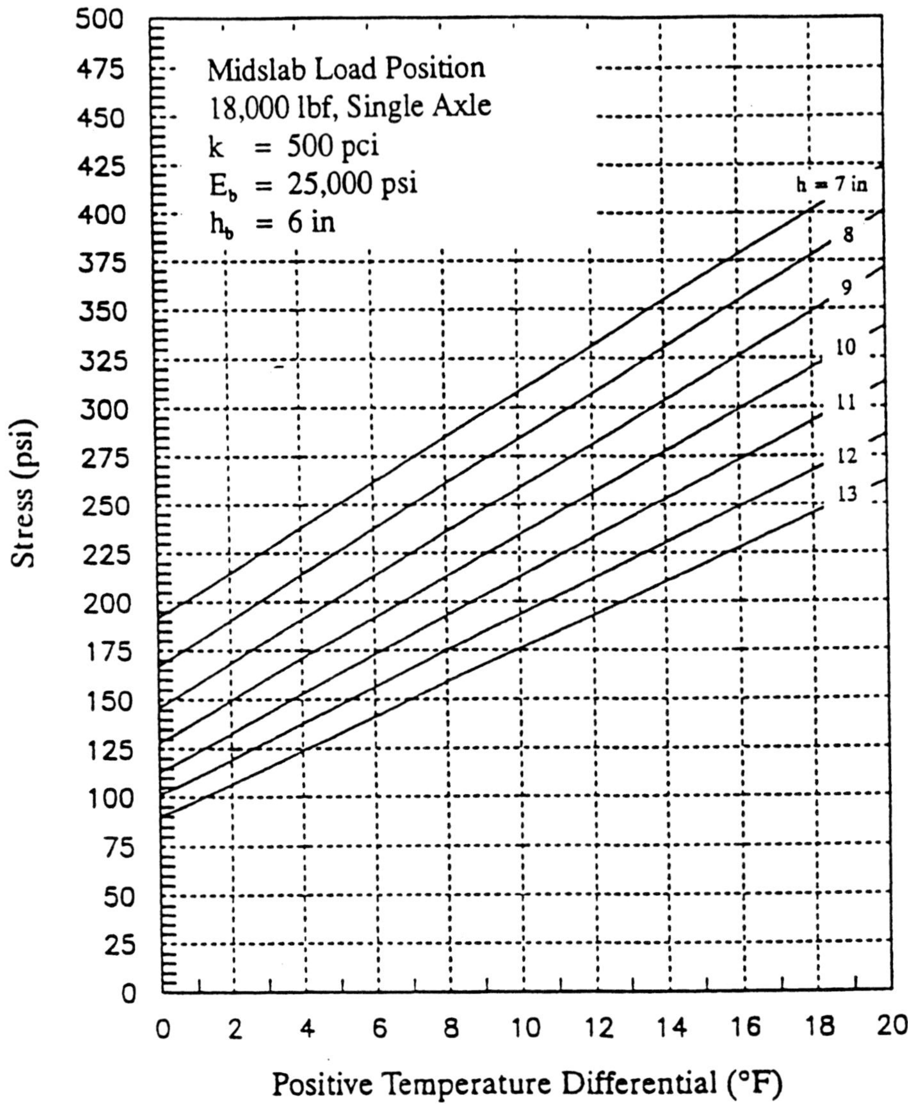

1 lbf = 4.45 N, 1 pci = 0.271 kPa/mm, 1 psi = 6.89 kPa, 1 in = 25.4 mm, °C = (°F - 32)/1.8

Figure 52. Tensile stress at bottom of slab for midslab loading position, positive temperature differential, and full friction, for aggregate base and stiff subgrade.

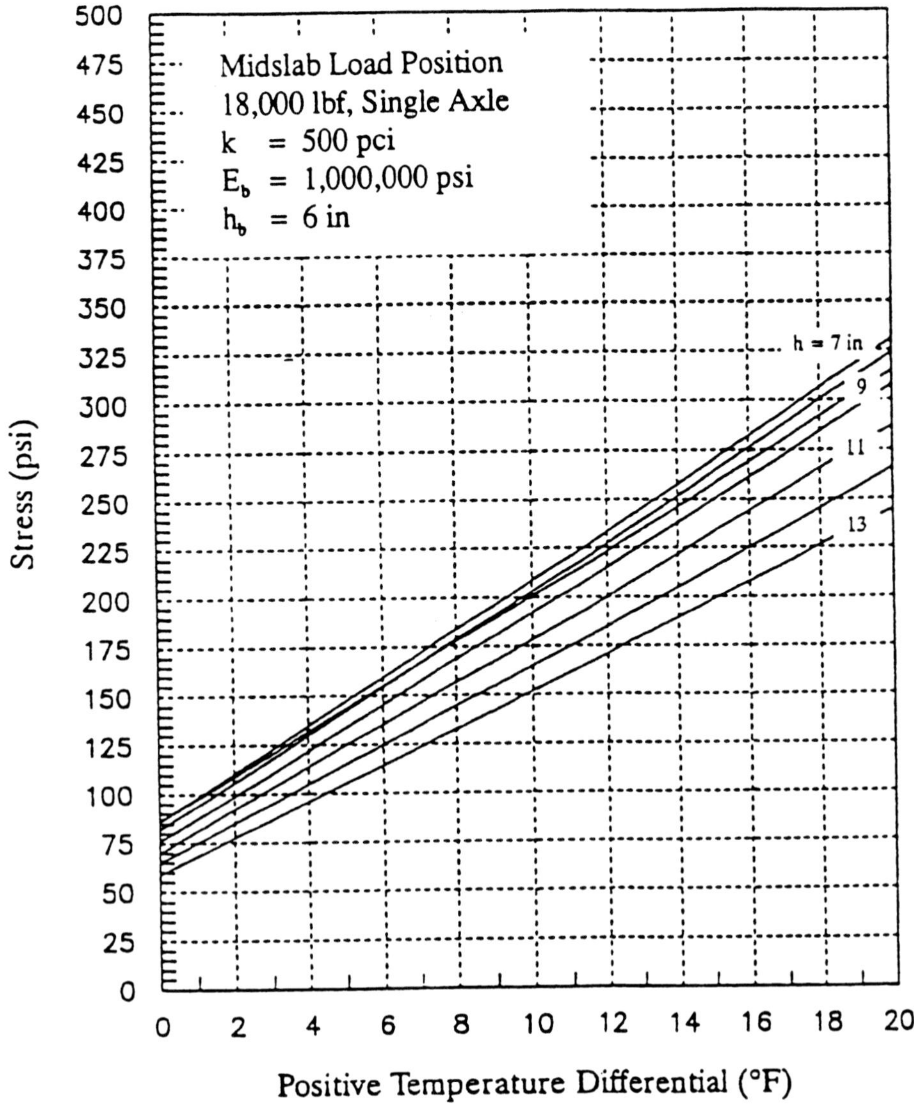

1 lbf = 4.45 N, 1 pci = 0.271 kPa/mm, 1 psi = 6.89 kPa, 1 in = 25.4 mm, °C = (°F - 32)/1.8

Figure 53. Tensile stress at bottom of slab for midslab loading position, positive temperature differential, and full friction, for high-strength base and stiff subgrade.

Step 4. Estimate the Critical Stress at the Top of the Slab From Joint Loading and Negative Temperature Differential using Figures 54 through 60. Charts are provided in Figures 54 through 59 for two levels of base modulus and three levels of subgrade. The full friction stress from Figures 54 through 59 is multiplied by the friction adjustment factor from Figure 60 to obtain the proper joint load stress.

Step 5: Compare the Midslab Load Position Stress at the Bottom of the Slab and the Joint Loading Position Stress at the Top of the Slab. If the joint load position produces a stress equal to or greater than the stress produced by midslab loading, strong consideration should be given to a redesign of the joints. Design features that protect against critical joint load stresses are the use of properly sized and spaced dowels, and, to a lesser degree, a widened slab (i.e., slab paved wider than 12 ft [3.66 m], but with the traffic lane striped 12 ft [3.66 m] wide) or tied concrete shoulder. The other effect that good load transfer has on performance is that corner deflections are reduced. High differential deflections can lead to erosion and loss of support, resulting in even greater stresses under corner loading. Reducing the joint spacing and/or changing the base type can also reduce stresses caused by joint loading.

Example Design Check for Joint Load Position Cracking. The adequacy of the joint design is checked in this example for the same design parameters as used previously in the example slab thickness determination. From the equations given previously, the tensile stress at the bottom of the slab due to midslab load and a positive temperature differential is calculated to be 188 psi [1296 kPa].

The negative effective temperature gradient is -5.3°F [-2.9°C]. The combined negative construction and moisture shrinkage is assumed for this example to be the maximum for a wet climate, -2°F/in [-0.044°C/mm] of slab thickness, or -22°F [-12°C]. Thus, the total negative temperature differential is about -27°F [-15°C]. Using Figures 54 through 59, the full friction tensile stress at the top of the slab due to joint load and negative curling is about 130 psi [896 kPa].

The full friction stress, when multiplied by the joint friction factor of 1.08 obtained from Figure 60, yields a joint loading stress of 140 psi [965 kPa]. The joint load position results in a lower stress than the midslab load position, so it is not necessary to modify the joint design to reduce the chance of corner breaks.

3.2.3 Stage Construction (no change)

3.2.4 Roadbed Swelling and Frost Heave (no change)

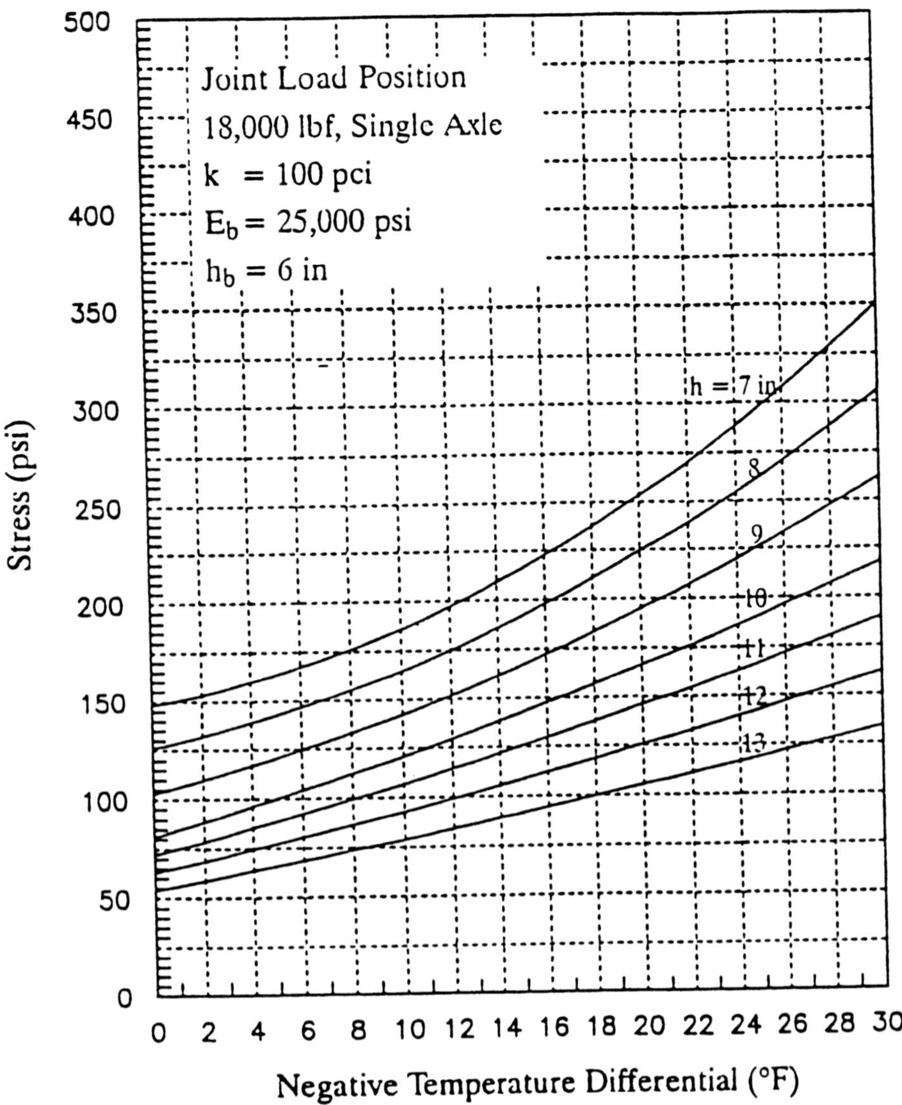

1 lbf = 4.45 N, 1 pci = 0.271 kPa/mm, 1 psi = 6.89 kPa, 1 in = 25.4 mm, °C = (°F - 32)/1.8

Figure 54. Tensile stress at top of slab for joint loading position, negative temperature differential, and full friction, for aggregate base and soft subgrade.

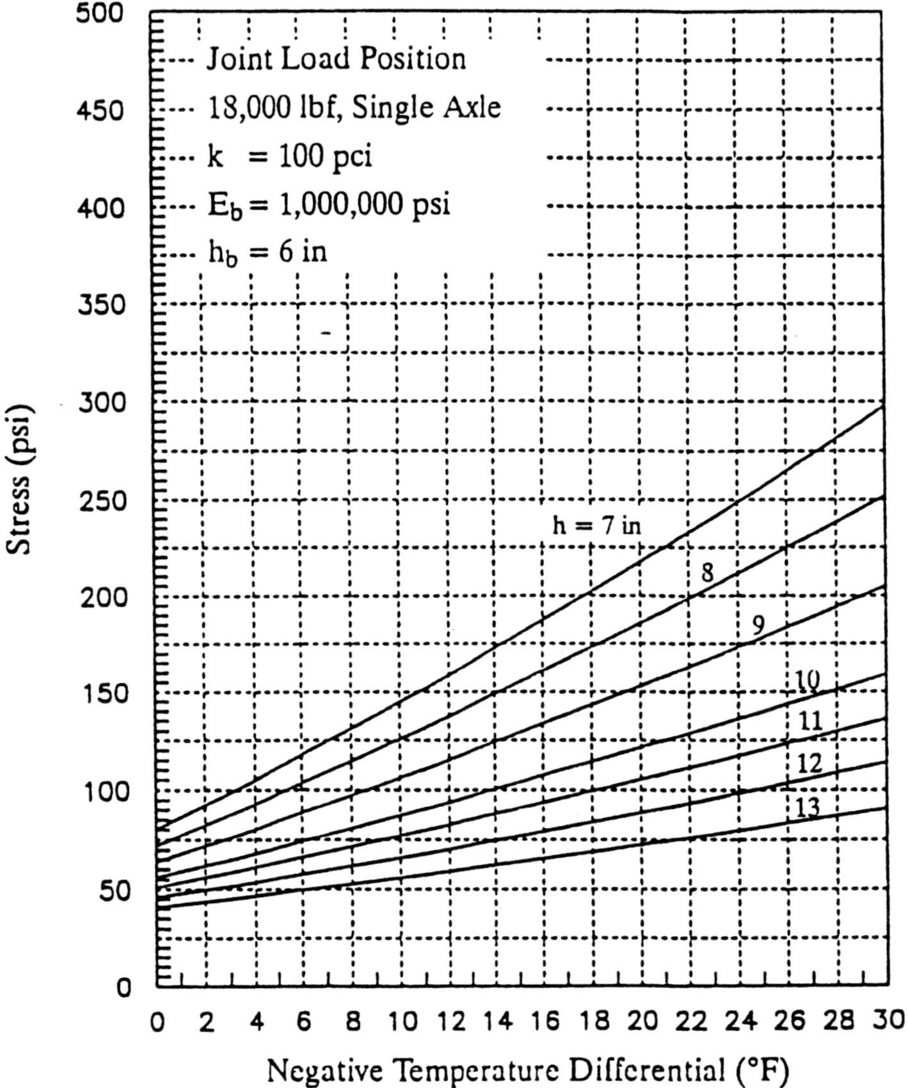

1 lbf = 4.45 N, 1 pci = 0.271 kPa/mm, 1 psi = 6.89 kPa, 1 in = 25.4 mm, °C = (°F - 32)/1.8

Figure 55. Tensile stress at top of slab for joint loading position, negative temperature differential, and full friction, for high-strength base and soft subgrade.

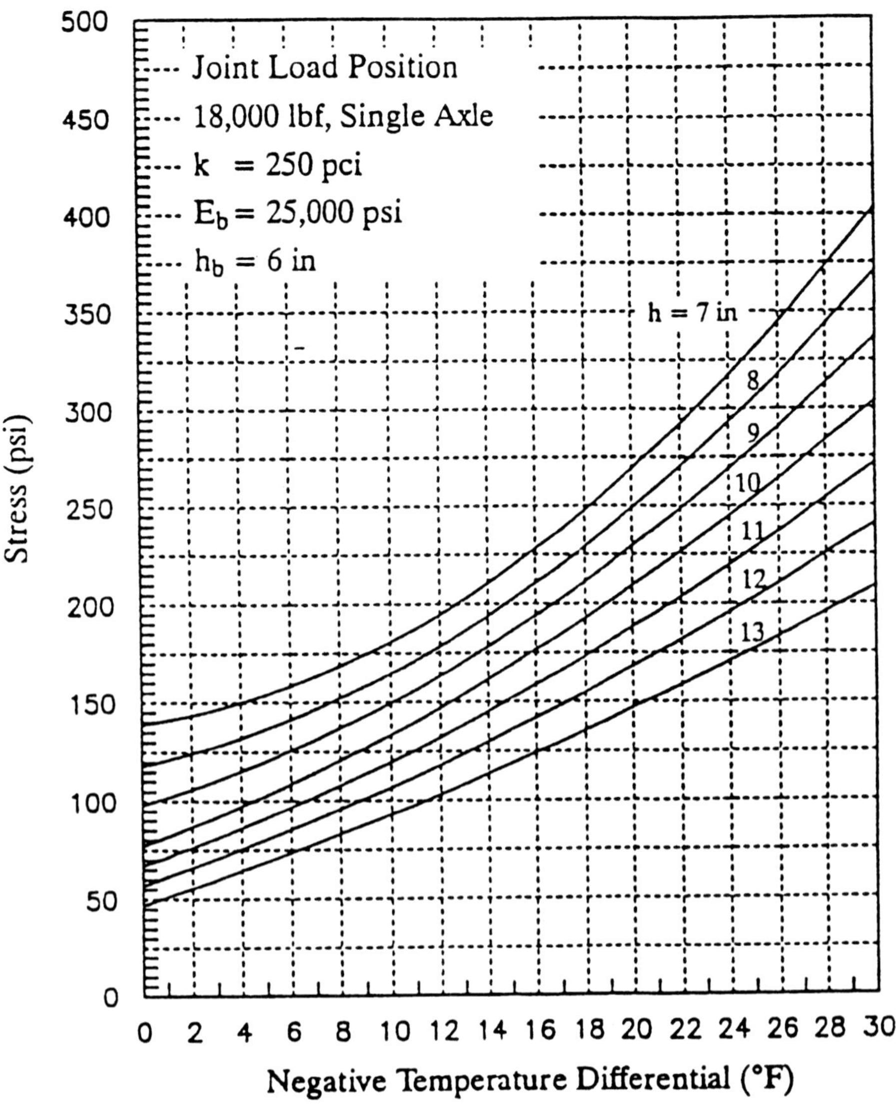

1 lbf = 4.45 N, 1 pci = 0.271 kPa/mm, 1 psi = 6.89 kPa, 1 in = 25.4 mm, °C = (°F - 32)/1.8

Figure 56. Tensile stress at top of slab for joint loading position, negative temperature differential, and full friction, for aggregate base and medium subgrade.

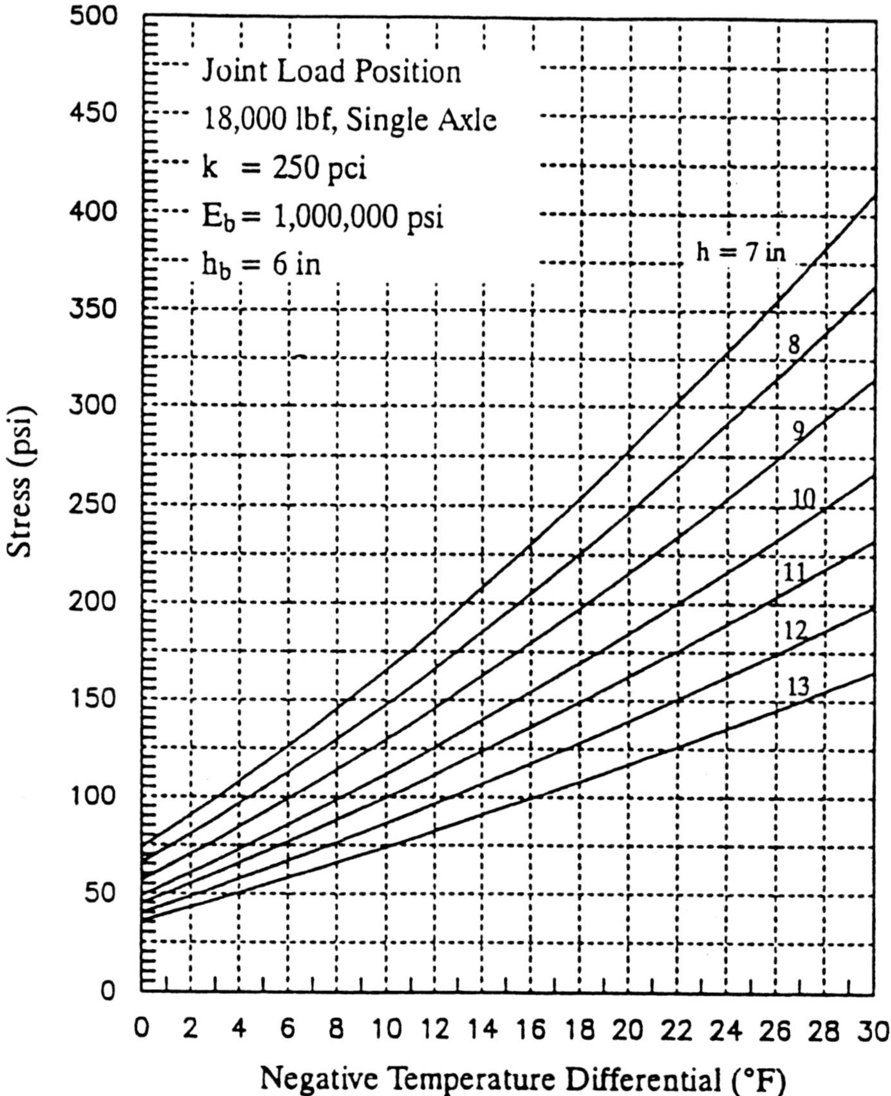

1 lbf = 4.45 N, 1 pci = 0.271 kPa/mm, 1 psi = 6.89 kPa, 1 in = 25.4 mm, °C = (°F - 32)/1.8

Figure 57. Tensile stress at top of slab for joint loading position, negative temperature differential, and full friction, for high-strength base and medium subgrade.

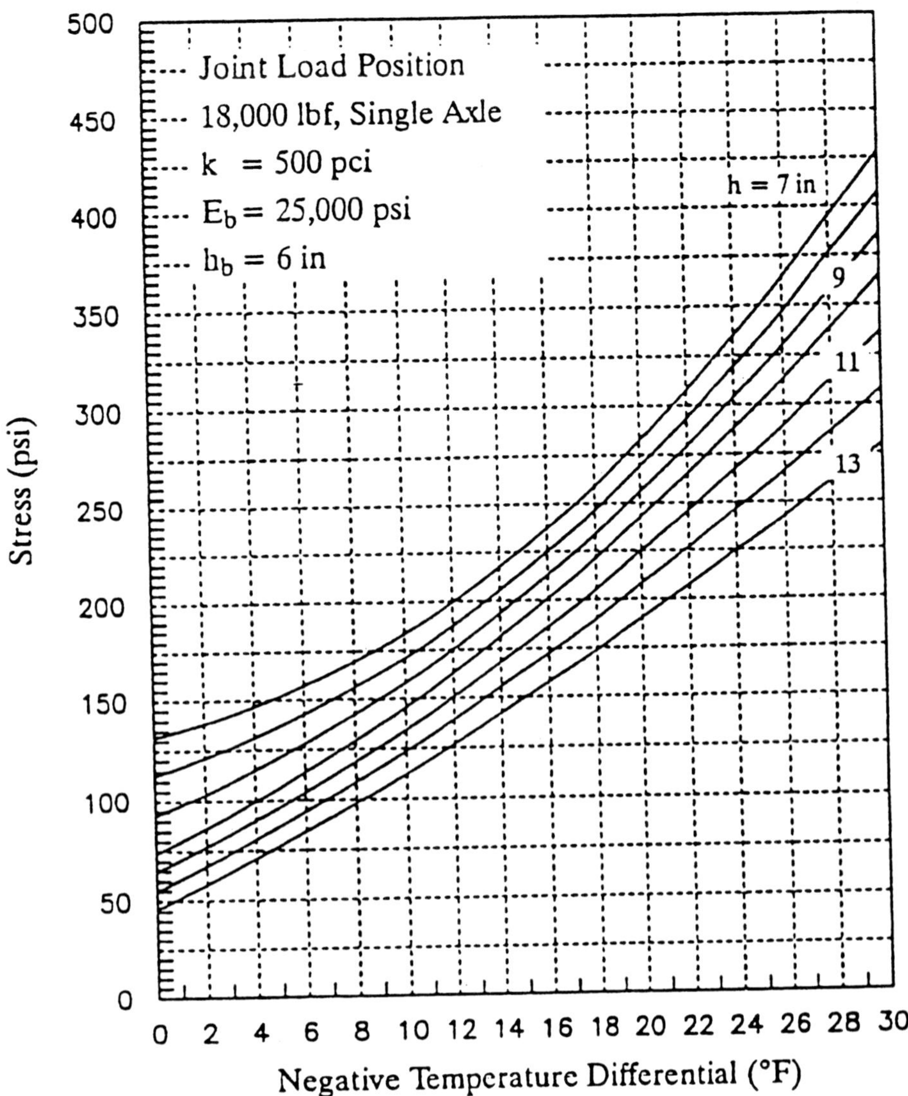

1 lbf = 4.45 N, 1 pci = 0.271 kPa/mm, 1 psi = 6.89 kPa, 1 in = 25.4 mm, °C = (°F - 32)/1.8

Figure 58. Tensile stress at top of slab for joint loading position, negative temperature differential, and full friction, for aggregate base and stiff subgrade.

59

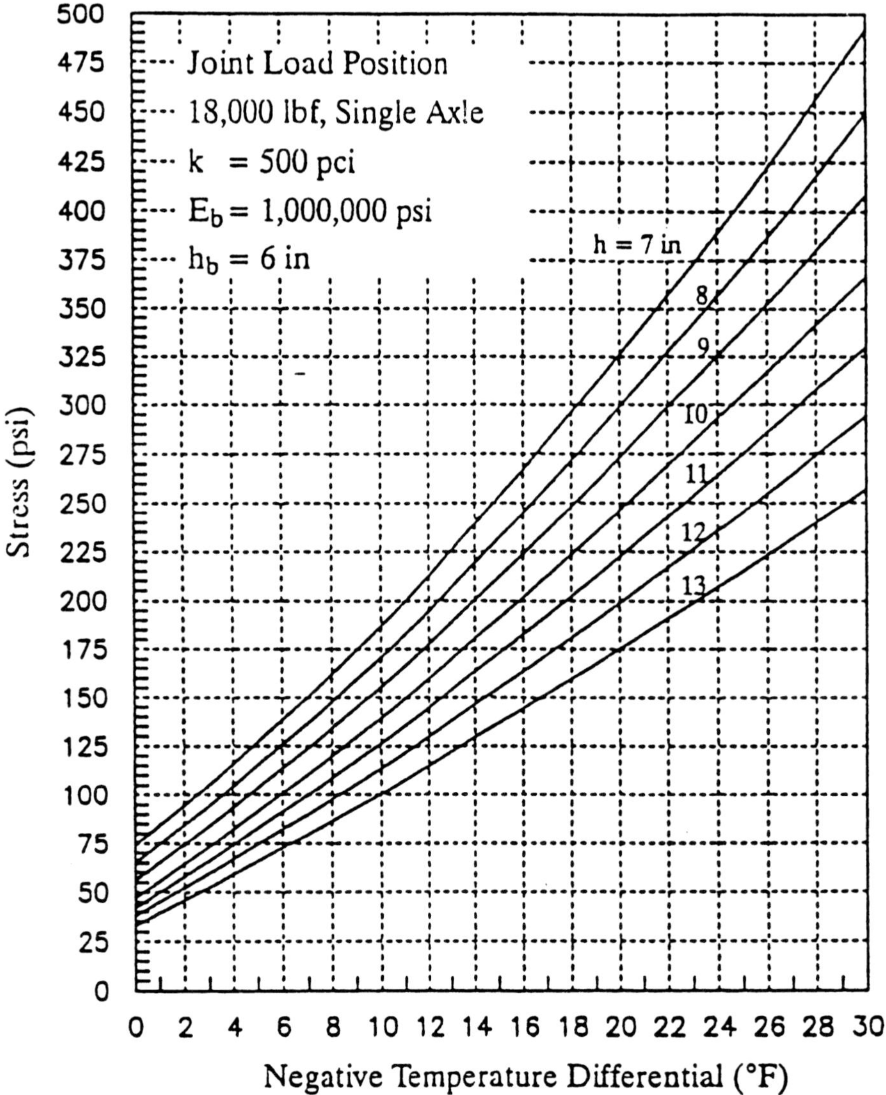

1 lbf = 4.45 N, 1 pci = 0.271 kPa/mm, 1 psi = 6.89 kPa, 1 in = 25.4 mm, °C = (°F - 32)/1.8

Figure 59. Tensile stress at top of slab for joint loading position, negative temperature differential, and full friction, for high-strength base and stiff subgrade.

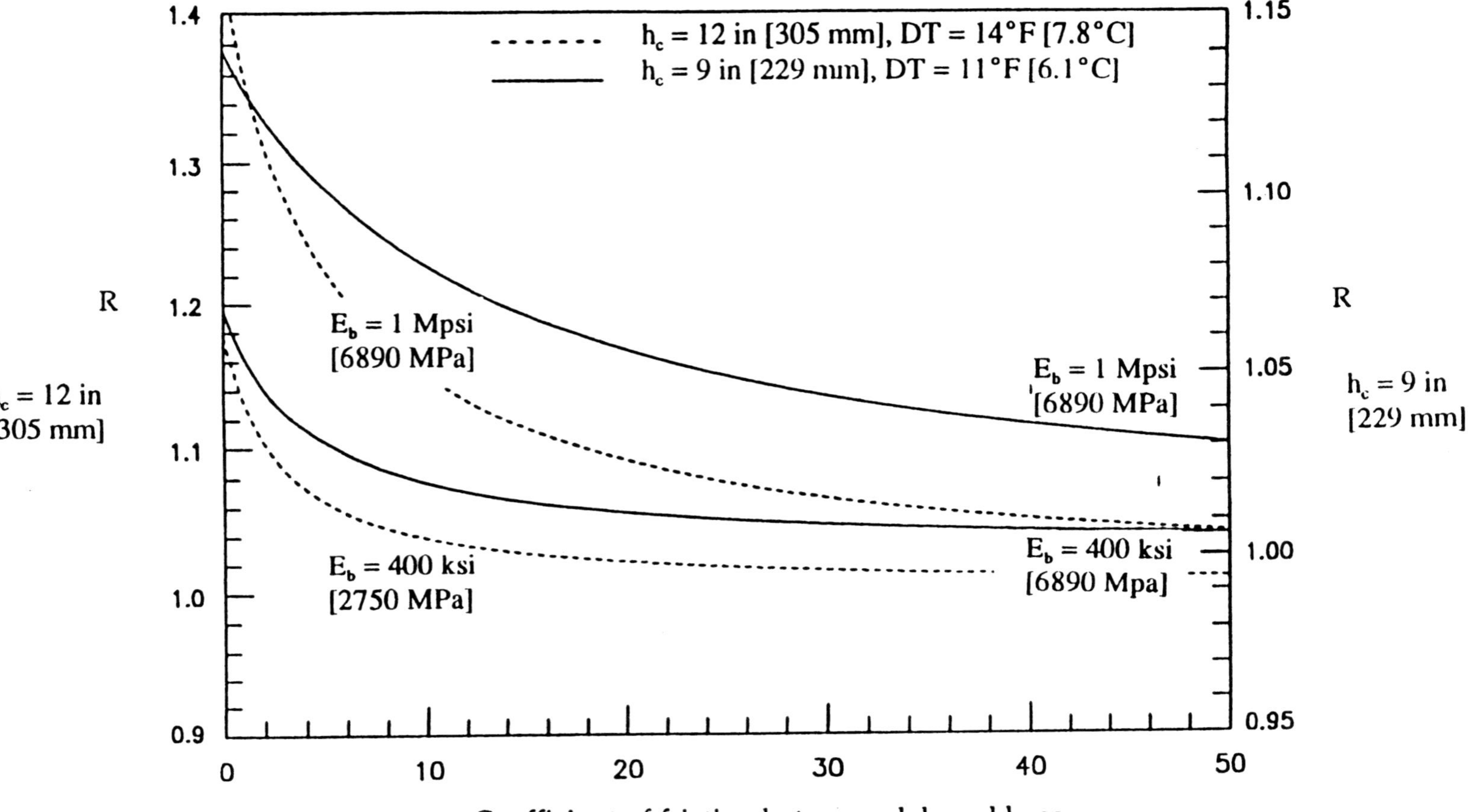

Note: $R = \sigma_{f} / \sigma_{full\,friction}$
hb = 5 in [127 mm], k = 200 psi/in [54 kPa/mm]
Joint loading with nighttime curling DT

Figure 60. Friction adjustment factor for stress at top of slab for joint loading.

3.3 RIGID PAVEMENT JOINT DESIGN

This section covers the design considerations for the different types of joints in portland cement concrete pavements. A joint faulting check is made after the required slab thickness is determined as described in Section 3.2.2.

3.3.1 Joint Types (no change)

3.3.2 Joint Geometry and Load Transfer

The joint geometry is considered in terms of the spacing, load transfer, and general layout.

Joint Spacing. In general, the spacing of both transverse and longitudinal contraction joints depends on local conditions of materials and environment, whereas expansion and construction joints are primarily dependent on layout and construction capabilities. For contraction joints, the spacing required to prevent intermediate cracking decreases as the thermal coefficient, positive temperature gradient, or base frictional resistance increases, and the spacing increases as the concrete tensile strength increases. Spacing is also related to the slab thickness and the joint sealant capabilities.

Determination of the required slab thickness includes an input for joint spacing. As joint spacing increases, stresses due to thermal curling and moisture warping increase. For JPCP and JRCP, the following recommendations are made.

JPCP (short-jointed plain concrete): Transverse cracking must be controlled. Increased joint spacing requires increased slab thickness, especially for stiffer bases and subgrades. The joint spacing interacts with slab thickness, base stiffness, subgrade stiffness (k-value), and also with the effective temperature gradient, which is location-dependent. Thus, there are tradeoffs between all of these variables that should be considered when selecting a design joint spacing. As a rough guide, the joint spacing (in feet) for plain concrete pavements should not exceed twice the slab thickness (in inches). For example, the maximum joint spacing for an 8-in [203-mm] slab is 16 ft [4.9 m]. For treated bases and stiff subgrades, this general guide may produce too long a joint spacing. Also, as a general guideline, the ratio of slab width to length should not exceed 1.25.

JRCP (long-jointed reinforced concrete): Transverse cracking is an expected occurrence and steel reinforcement is provided to hold the cracks tight. For JRCP, the designer should input a joint (crack) spacing of 30 ft [9.1 m] for thickness design purposes only.

For both JPCP and JRCP, local performance data are valuable for helping to establish a joint spacing that will control cracking. Local experience must be tempered since a change in any of several concrete properties or construction methods (e.g., a change in coarse aggregate type), may have a significant impact on the concrete thermal coefficient and, consequently, the acceptable joint spacing.

The use of expansion joints is generally minimized on a project due to cost, complexity, and performance problems. They are used at structures where pavement types change (e.g., CRCP to jointed), with prestressed pavements, and at intersections.

The spacing between construction joints is generally dictated by field placement and equipment capabilities. Longitudinal construction joints should be placed at lane edges to maximize pavement smoothness and minimize load transfer problems. Transverse construction joints occur at the end of a day's placement or in connection with equipment breakdowns.

Joint Load Transfer. Because the joints of the AASHO Road Test pavements were adequately doweled, no significant faulting occurred during the 2 years of the experiment. If the joints had not been properly doweled, substantial faulting would have occurred, which would have greatly changed the rigid pavement performance model.

Faulting is one of the most important distresses affecting rideability and serviceability. A pavement that faults significantly will have reduced serviceability and carry fewer traffic loads to terminal serviceability than a pavement of the same cross section that does not fault. Thus, joints must be prevented from significant faulting through good joint load transfer and spacing design, base design, and subdrainage design.

The procedure to check the adequacy of the proposed joint load transfer design consists of the following steps:

1. Determine the required slab thickness as described previously (including, if the pavement will be undoweled, the check to compare midslab loading stress to joint loading stress).

2. Predict the mean joint faulting using the appropriate model for doweled or undoweled pavements.

3. Compare the predicted mean faulting to the critical faulting level recommended to prevent faulting from contributing significantly to serviceability loss. If the predicted mean faulting exceeds the critical level, the joint load transfer design should be modified.

Step 1. Determine the Required Slab Thickness. For undoweled pavement, the check for cracking due to joint loading is conducted as well. The joint design features may be modified if necessary and a redesign made to achieve an acceptable joint design to prevent cracking.

Step 2. Predict the Mean Joint Faulting Over the Design Life using the faulting prediction models given below.

Faulting Model for Doweled Joints:

$$\text{FaultD} = \text{CESAL}^{0.25} * [0.0628 - 0.0628 * C_d + 0.3673*10^{-8} * \text{Bstress}^2$$
$$+ 0.4116 * 10^{-5} * \text{Jtspace}^2 + 0.7466*10^{-9} * \text{FI}^2 * \text{Precip}^{0.5} \qquad [52]$$
$$- 0.009503 * \text{Basetype} - 0.01917 * \text{Widenlane} + 0.0009217 * \text{Age}]$$

where FaultD = mean transverse doweled joint faulting, inches

CESAL = cumulative equivalent 18-kip [80-kN] single-axle loads, millions

C_d = modified AASHTO drainage coefficient:

Edge Drains	Precip. Level	Fine-Grained Subgrade		Coarse-Grained Subgrade	
		Nonpermeable Base	Permeable Base	Nonpermeable Base	Permeable Base
No	Wet	0.70-0.90	0.85-0.95	0.75-0.95	0.90-1.00
	Dry	0.90-1.10	0.95-1.10	0.90-1.15	1.00-1.15
Yes	Wet	0.75-0.95	1.00-1.10	0.90-1.10	1.05-1.15
	Dry	0.95-1.15	1.10-1.20	1.10-1.20	1.15-1.20

Notes:
1. Fine subgrade = A-1 through A-3 classes;
   Coarse subgrade = A-4 through A-8 classes.
2. Permeable Base = k = 1000 ft/day (305 m/day) or uniformity coefficient (C_u) ≤ 6.
3. Wet climate = Precipitation > 25 in/year (635 mm/year);
   Dry climate = Precipitation ≤ 25 in/year (635 mm/year).
4. Select midpoint of range and use other drainage features (adequacy of cross slopes, depth of ditches, presence of daylighting, relative drainability of base course, bathtub design, etc.) to adjust upward or downward.

BSTRESS = maximum concrete bearing stress from closed-form equation, psi:

$$BSTRESS = f_d \, P \, T \left[\frac{K_d \, (2 + BETA * OPENING)}{4 \, E_s \, I \, BETA^3} \right] \qquad [53]$$

$$BETA = \sqrt[4]{\frac{K_d \, DOWEL}{4 \, E_s \, I}} \qquad [54]$$

f_d = distribution factor = $2 * 12 / (\ell + 12)$

ℓ = radius of relative stiffness, inches

I = moment of inertia of dowel bar cross section, in^4 :

$$I = 0.25 \, \pi \left(\frac{DOWEL}{2} \right)^4 \tag{55}$$

P = applied wheel load, set to 9000 lbf [40 kN]

T = percent transferred load, set to 0.45

K_d = modulus of dowel support, set to 1,500,000 psi/in [405 MPa/mm]

BETA = relative stiffness of the dowel-concrete system

DOWEL = dowel diameter, inches

E_s = modulus of elasticity of the dowel bar, psi

k = modulus of subgrade reaction, psi/in

OPENING = average transverse joint opening, inches:

$$OPENING = 12 * CON * Jtspace * \left(\frac{ALPHA * TRANGE}{2 + e} \right) \tag{56}$$

Jtspace = average transverse joint spacing, ft

CON = adjustment factor due to base/slab frictional restraint:

= 0.65 if stabilized base

= 0.80 if aggregate base or lean concrete base with bond breaker

ALPHA = PCC thermal expansion coefficient, set to 0.000006/°F [0.000003/°C]

TRANGE = annual temperature range, °F

e = PCC drying shrinkage coefficient, set to 0.00015 strain

FI = mean annual freezing index, Fahrenheit degree-days

Precip = mean annual precipitation, inches

Basetype =   0 for unstabilized base, 1 for stabilized base

Widenlane =   0 if not widened, 1 if widened

Age =   pavement age, years

Faulting Model for Undoweled Joints:

$$\text{FautND} = \text{CESAL}^{0.25} * [0.2347 - 0.1516 * C_d - 0.000250 * \text{Slabthick}^2/\text{Jtspace}^{0.25}$$
$$-0.0155 * \text{Basetype} + 0.7784*10^{-7} * \text{FI}^{1.5} * \text{Precip}^{0.25} \qquad [57]$$
$$-0.002478 * \text{Days90}^{0.5} - 0.0415 * \text{Widenlane}]$$

where FaultND =   mean transverse undoweled joint faulting, inches

Days90 =   number of days with maximum temperature above 90°F [32.2°C]

and all other variables are as defined for FaultD.

Tables 25, 26, and 27 were developed as examples using the above equations to show the faulting predictions for pavements with and without dowel bars, for the design parameters shown.

Table 25. Mean joint faulting predictions for doweled jointed plain concrete pavement using Equation 52.

ESALs, millions	Granular Base			Treated Base		
	Dowel Diameter 1.00 in	Dowel Diameter 1.25 in	Dowel Diameter 1.50 in	Dowel Diameter 1.00 in	Dowel Diameter 1.25 in	Dowel Diameter 1.50 in
1	0.03	0.01	0.01	0.02	0.00	0.00
2.5	0.04	0.02	0.01	0.03	0.01	0.00
5	0.05	0.03	0.02	0.04	0.01	0.00
10	0.07	0.04	0.03	0.05	0.02	0.01
20	0.10	0.07	0.06	0.08	0.05	0.04
30	0.13	0.10	0.08	0.11	0.07	0.06
40	0.16	0.13	0.11	0.14	0.10	0.09
50	0.20	0.16	0.14	0.17	0.13	0.12
75	0.29	0.24	0.23	0.26	0.22	0.20
100	0.38	0.33	0.32	0.35	0.30	0.29

Values shown in table are mean predicted joint faulting, inches [1 in = 25.4 mm]

Joint spacing = 15 ft [4.6 m]
k-value = 100 psi/in [27 kPa/mm]
Precipitation = 30 in/year [762 mm/year]
FI = 200°F [93.3°C]-days
Lane not widened

Slab thickness = 9 in [229 mm]
E = 4,000,000 psi [27,580 MPa]
TRANGE = 85°F [29.4°C] (July max - January min)
E_s = 29,000,000 psi [200,000 MPa]
Age = ESALs in millions

Table 26. Mean joint faulting predictions for doweled jointed reinforced concrete pavement using Equation 52.

ESALs, millions	Granular Base			Treated Base		
	Dowel Diameter 1.00 in	Dowel Diameter 1.25 in	Dowel Diameter 1.50 in	Dowel Diameter 1.00 in	Dowel Diameter 1.25 in	Dowel Diameter 1.50 in
1	0.04	0.02	0.02	0.03	0.01	0.01
2.5	0.05	0.03	0.02	0.04	0.02	0.01
5	0.06	0.04	0.03	0.05	0.02	0.02
10	0.08	0.05	0.04	0.07	0.04	0.03
20	0.12	0.08	0.07	0.10	0.06	0.05
30	0.15	0.12	0.10	0.13	0.09	0.08
40	0.19	0.15	0.13	0.16	0.12	0.11
50	0.22	0.18	0.16	0.20	0.15	0.14
75	0.31	0.27	0.25	0.28	0.24	0.22
100	0.41	0.36	0.34	0.38	0.33	0.31

Values shown in table are mean predicted joint faulting, inches [1 in = 25.4 mm]

Joint spacing = 45 ft [13.7 m]
k-value = 100 psi/in [27 kPa/mm]
Precipitation = 30 in/year [762 mm/yr]
FI = 200°C [93.3°C]-days
Lane not widened

Slab thickness = 9 in [229 mm]
E = 4,000,000 psi [27,580 MPa]
TRANGE = 85°F [29.4°C] (July max - January min)
E_s = 29,000,000 psi [200,000 MPa]
Age = ESALs in millions

Table 27. Mean joint faulting predictions for undoweled jointed plain concrete
pavement using Equation 57.

ESAL, millions	$C_d = 0.80$		$C_d = 1.0$	
	Joint Spacing 15 ft	Joint Spacing 20 ft	Joint Spacing 15 ft	Joint Spacing 20 ft
1	0.09	0.09	0.05	0.05
2.5	0.12	0.12	0.06	0.06
5	0.14	0.14	0.08	0.08
10	0.16	0.17	0.09	0.09
20	0.20	0.20	0.11	0.11
30	0.22	0.23	0.12	0.12
40	0.23	0.23	0.13	0.13
50	0.25	0.25	0.13	0.14
75	0.27	0.27	0.15	0.15
100	0.29	0.29	0.16	0.16

Values shown in table are mean predicted joint faulting, inches [1 in = 25.4 mm]

Joint spacing = 15 or 20 ft [4.6 or 6.1 m]
Slab thickness = 9 in [229 mm]
Precipitation = 30 in/year [762 mm/year]
FI = 200°F [93.3°C]-days
Days90 = 20
Lane not widened

Step 3. Compare the Predicted Mean Faulting With the Recommended Maximum Critical Levels given in Table 28. If the predicted faulting is greater than the recommended level, an adjustment to the joint load transfer design should be made. Potential adjustments include use of dowels, or, if dowels already exist, an increase in the diameter; selection of a different base type and permeability; and/or a decrease in the joint spacing (for undoweled joints).

Slab thickness should not be increased in an effort to improve the joint load transfer design, because slab thickness has only a minimal effect on joint faulting. However, the slab design may need adjustment after the joint design is completed, especially if the joint spacing is reduced or the base type is changed to reduce expected faulting.

Table 28. Recommended critical mean joint faulting levels for design.

Joint Spacing	Critical Mean Joint Faulting
Less than 25 ft	0.06 in
Greater than 25 ft	0.13 in

1 ft = 0.305 m, 1 in = 25.4 mm

These critical levels were derived from analysis of extensive field data. The mean faulting was computed for pavements with a serviceability of 3.0 or less. For example, based upon data from many short-jointed JPCP sections, a mean joint faulting of 0.12 in [3 mm] corresponded to a serviceability index of 3.0 or less. For long-jointed JRCP, the mean faulting level was 0.26 in [6.6 mm]. The recommended critical levels for design were selected as 50 percent of these values in order to effectively exclude faulting as a significant contributor to serviceability loss.

Example check for joint faulting. Assume the same pavement defined in the previous examples. The pavement has a 16-ft [4.9-m] joint spacing, treated base, subdrains, and no dowel bars. A Freezing Index of 500°F [260°C]-days, an annual temperature range of 85°F [47.2°C], and an annual precipitation of 30 in [762 mm] are also assumed for the location. A slab thickness of 10.75 in [273 mm] was obtained for a design traffic of 20 million ESALs and 95 percent reliability. The mean predicted joint faulting of 0.13 in [3.3 mm] exceeds the recommended limit of 0.06 in [1.5 mm], and thus the joint design is inadequate. One possible design modification would be to specify 1.25-in-diameter [32-mm-diameter] dowels. The mean predicted joint faulting would then be 0.05 in [1.27 mm], which would be acceptable.

Joint Layout (no change)

Joint Dimensions (no change)

3.3.3 Joint Sealant Dimensions (no change)

RIGID PAVEMENT DESIGN EXAMPLE

(PROPOSED REVISION TO AASHTO GUIDE APPENDIX I)

A jointed concrete pavement is to be designed to carry 10 million ESALS and the pavement is located in the southeastern United States.

<u>GENERAL DESIGN INPUTS</u>

Design reliability = 90 percent

Overall standard deviation, S_0 = 0.39

Design traffic = 10 million ESALs in the design lane

P1 - P2 = 4.5 - 2.5 = 2.0

Concrete flexural strength, mean 28-day, third-point loading, S'_c = 700 psi [4827 kPa]

Concrete elastic modulus, E_c = 4,100,000 psi [28,270 MPa]

Subgrade soil type: silty clay

k-value = elastic value of subgrade/embankment = 100 psi/in [27 kPa/mm]

Subdrains = 1 (yes)

Climate: WIND =   mean annual wind speed = 7.9 mph [12.7 km/h]
           TEMP =   annual temperature = 58.9°F [14.9°C]
           PREC =   annual precipitation = 43 in [1092 mm]

Effective positive temperature differential:

Slab Thickness	Temperature Differential
9 in [229 mm]	8.3°F [4.6°C]
10 in [254 mm]	8.9°F [4.9°C]
11 in [279 mm]	9.4°F [5.2°C]

Freezing Index = 0°F [0°C]-days below freezing

Temperature Range = 50°F [27.7°C] (maximum July - minimum January)

DESIGN ALTERNATIVE A

Undoweled joints

Untreated aggregate base, 6 in [152 mm], E_b = 25,000 psi [172 MPa], friction f = 1.5

Joint spacing = 15 ft [4.6 m]

Conventional lane width = 12 ft [3.7 m]

AC shoulders

Slab Thickness Design
Assuming an effective temperature differential of about 9°F [5°C], a required slab thickness of 10.2 in [259 mm] is obtained for design ESALs of 10 million, at a design reliability level of 90 percent.

Joint Faulting Check
The initial design has undoweled joints with a 15-ft [4.6-m] joint spacing. The estimated mean faulting for this design is 0.09 in [2.3 mm]. This value exceeds the recommended limit of 0.06 in [1.5 mm]. Therefore, a joint design modification (e.g., dowels, shorter joint spacing, different base type, tied shoulder) is required to control faulting.

Joint Load Position Stress Check
The joint load position check is required since the pavement is undoweled. The total negative temperature differential is estimated from the climatic data as -5.6°F [-3.11°C] (use -6°F [-3.33°C]).

Combined moisture gradient and construction differential: -10°F [-5.6°C] (wet climatic zone, conventional concrete cure).

Total negative equivalent temperature differential: -16°F [-8.89°C].

The critical stress for joint loading is determined to be about 145 psi [1000 kPa] for a slab thickness of 10.2 in [259 mm]. This joint loading stress is compared to that obtained for the midslab location with a positive temperature differential of 9°F [5°C], which is found to be 233 psi [1607 kPa]. Therefore, the midslab load design is adequate to control stresses at the joint loading position. A total negative temperature differential of about -30°F [-16.67°C] would be required to produce a stress greater than 233 psi [1607 kPa].

DESIGN ALTERNATIVE B

Undoweled joints

Permeable asphalt-treated aggregate base, 6 in [152 mm], E_b = 100,000 psi [690 MPa], friction f = 6

Joint spacing = 15 ft [4.6 m]

Widened slab width = 14 ft [4.3 m] (with AC shoulders)

Slab Thickness Design
Assuming an effective positive temperature differential of about 9°F [5°C], a required slab thickness of 9.4 in [239 mm] is obtained. Note that a stress reduction factor of 0.92 for the widened slab was used in the calculation.

Joint Faulting Check
The mean faulting estimated for this design is 0.06 in [1.5 mm], which just equals the recommended limit. Therefore, the joint design is acceptable.

Joint Load Position Stress Check
The joint load position check is required since the pavement is undoweled. The total negative temperature differential is the same as estimated for Alternative A, -16°F [-8.89°C].

The critical stress for joint loading is determined to be 165 psi [1138 kPa] for a slab thickness of 9.4 in [239 mm]. This stress is compared to that obtained for the midslab location with a positive temperature differential of 9°F [5°C], which is found to be 234 psi [1613 kPa]. Therefore, the midslab load design is adequate to control stresses at the joint loading position.

DESIGN ALTERNATIVE C

Doweled joints, 1.25 in [32 mm] diameter

Untreated aggregate base, 6 in [152 mm], E_b = 25,000 psi [172 MPa], friction f = 1.5

Joint spacing = 17 ft [5.2 m]

Conventional lane width = 12 ft [3.7 m]

Tied concrete shoulder

Slab Thickness Design
Assuming an effective temperature differential of about 9°F [5.0°C], the required slab thickness is 9.9 in [251 mm]. Note that a stress reduction factor of 0.94 for a tied concrete shoulder was used in the calculation.

<u>Joint Faulting Check</u>
The estimated mean faulting for this design is 0.01 in [0.25 mm], which is well below the 0.06-in [1.5-mm] recommended limit.

<u>Joint Load Position Stress Check</u>
The joint load position check is not required since the pavement is doweled and the joint load position stress will be well below the midslab stress.